A.I.
HACKED

A.I. HACKED

A PRACTICAL GUIDE TO THE FUTURE WITH ARTIFICIAL INTELLIGENCE

ELZAR SIMON

Copyright © 2019 Elzar Simon.

All rights reserved. No part of this book may be used or reproduced by any means, graphic, electronic, or mechanical, including photocopying, recording, taping or by any information storage retrieval system without the written permission of the author except in the case of brief quotations embodied in critical articles and reviews.

This book is a work of non-fiction. Unless otherwise noted, the author and the publisher make no explicit guarantees as to the accuracy of the information contained in this book and in some cases, names of people and places have been altered to protect their privacy.

Archway Publishing books may be ordered through booksellers or by contacting:

Archway Publishing
1663 Liberty Drive
Bloomington, IN 47403
www.archwaypublishing.com
1 (888) 242-5904

Because of the dynamic nature of the Internet, any web addresses or links contained in this book may have changed since publication and may no longer be valid. The views expressed in this work are solely those of the author and do not necessarily reflect the views of the publisher, and the publisher hereby disclaims any responsibility for them.

Any people depicted in stock imagery provided by Getty Images are models, and such images are being used for illustrative purposes only.
Certain stock imagery © Getty Images.

ISBN: 978-1-4808-7835-8 (sc)
ISBN: 978-1-4808-7834-1 (hc)
ISBN: 978-1-4808-7836-5 (e)

Library of Congress Control Number: 2019907257

Print information available on the last page.

Archway Publishing rev. date: 10/24/2019

This book is dedicated to my lovely wife, Elisa, and to our wonderful children, Joshua, Elisha, and Ysabel.

My special thanks to Leigh Berrell, Allan Jansen, Natalija Jovanovic, and Maurice Winn.

CONTENTS

Introduction		ix
CHAPTER 1	ANALOG TO DIGITAL LIFE	1
CHAPTER 2	THE BIRTH OF ARTIFICIAL INTELLIGENCE	5
CHAPTER 3	VALUE PROPOSITION	11
	Generating Profit	12
	Business Growth	14
CHAPTER 4	HOMEWARD BOUND	17
CHAPTER 5	THE THINGS THAT MAKE US HUMAN	21
CHAPTER 6	EMOTIONAL INTELLIGENCE	25
CHAPTER 7	THE ART OF MAKING ART	29
	AI in Music	30
	AI in Fine Arts and Literature	32
	Where the Real Value Lies	32
CHAPTER 8	NOW THAT'S ENTERTAINMENT!	37
	Creating and Marketing Content	37
	The Ultimate Virtual Experience	39
CHAPTER 9	MINDING MY OWN BUSINESS	41
	The Matter of Privacy	41
	Equal and Opposite Reaction	43
	Body Chip	45

CHAPTER 10	WHAT ABOUT MY JOB?	49
CHAPTER 11	TEACH YOUR CHILDREN	55
	Personalized Learning	55
	Refocusing Education	58
CHAPTER 12	TALK TO ME	61
	Text-based Communication	61
	Verbal Communication	62
	AI Enhanced Devices and Apps	64
CHAPTER 13	YOU GOT A FRIEND IN ME	67
CHAPTER 14	REMEMBER MASLOW	71
CHAPTER 15	THE SCIENTIFIC WAY	77
	Accelerating Drug Discovery	77
	Smart Medical Devices	80
	AI in Science and Technology	82
CHAPTER 16	JUSTICE DELAYED IS JUSTICE DENIED	85
CHAPTER 17	A MIGHTY FORTRESS	89
CHAPTER 18	WHAT MAKES THE WORLD GO 'ROUND	91
CHAPTER 19	THE NEXT FRONTIER	95
Index		101
Notes		105

INTRODUCTION

Like a tumultuous torrent of freshwater frenzy rushing through the mighty Amazon River, an insuppressible movement is sweeping throughout the world today.

This silent pervasion suddenly gained media attention in early 2011 at an event that was wrapped with entertainment: IBM Watson[1] beat highly intelligent human beings on *Jeopardy*, a famous television game show.

Jeopardy[2] is a game of wit. Those who are familiar with the TV show know that the subjects are broad and unpredictable, the topics vary from the very trivial to common knowledge, the questions are phrased like puzzles, and the answers must be given by contestants in the form of a question. IBM Watson's victory was like a checkered flag that was raised to signal the start of a historic race.

Since then, the ebullient revolution has forged ahead.

In January 2016, the *Telegraph* released an article entitled "First driverless buses travel public roads in Netherlands."[3]

In April 2016, *MIT Technology Review* released an article entitled "Will Artificial Intelligence Win the Caption Contest?" The same article heralded, "Neural networks have mastered the ability to label things in images, and now they're learning to tell stories from a set of photos."[4]

In October 2016, *Forbes* (forbes.com) released an article entitled "The World's First Home Robotic Chef Can Cook over 100 Meals."[5]

In January 2017, *Quartz* (qz.com), a news provider covering the global economy, came out with an article entitled "Japanese White-Collar Workers Are Already Being Replaced by Artificial Intelligence."[6]

In June 2017, an article in *Bloomberg* announced that PricewaterhouseCoopers (PwC), which is one of the "Big Four" consulting companies in the world, predicted that "the global GDP would be 14% higher in 2030 as a result of Artificial Intelligence."[7] That's in 2030—just a few years from now.

In October 2018, Christie's Auction House (www.christies.com), a world-famous auction house in New York City, sold an AI-generated artwork at a price that was forty times higher than what the sellers expected.[8]

At the ground level, the movement is conspicuous. We now post a picture on Facebook[9] and the system can recognize us and our friends in it. We can talk to a home device like the Amazon Alexa[10] or Google Home[11] and ask it to play us a song, turn off our television set, tell us about the weather, or converse on any topic under the sun—in exactly the same way that we talk to humans. We read about Tesla's technological breakthrough in manufacturing cars that could drive us to our destination without any human intervention.[12] We order an item on Amazon Prime,[13] and the shopping site suggests a list of related products that other people bought when they purchased the item we just selected. We visit a bank website and a chatbot[14] pops up to ask us if we have any questions.

Artificial intelligence (AI) has arrived and will continue to be a forceful reality that we humans have to reckon with—sooner than we might think.

AI is here and we have a choice: we can either ignore or acknowledge its existence.

To dismiss its existence would be imprudent, like an ostrich that buries its head in the sand.

While intelligent machines would naturally have human cynics and skeptics, AI would unquestionably play a major role in our new world order, whether we like it or not. In fact, AI's building blocks are now in place: machine perception,[15] facial and voice recognition,[16] deep learning or machine learning,[17] and of course, natural language processing[18] that enables machines to audibly or textually converse with us humans. And let's not forget sophisticated gaming algorithms, advanced AI planning capabilities, and robotics.[19]

Regardless of ethics, we humans would endeavor to create anything that could be made and explore anything that could be discovered. That is just part of our DNA[20] as a human race. And that is the key factor as to why we humans have made significant progress in many fields over the past thousands of years. And we would assuredly continue to do so.

While artificial intelligence is perceived by many as the ultimate enemy of humankind, many envision that AI would help solve humanity's biggest challenges: health care, medical research, food production, transportation, security and cybercrime, and many more. As AI algorithms become more sophisticated, this unstoppable force will continue to advance across humankind's dominion. There is no stopping this revolution.

It is therefore imperative that we vigilantly watch its development, judiciously examine its impact, and actively participate in factual and pragmatic deliberations.

What is AI really capable of doing?

Would these intelligent machines simply complement the human workforce or replace us?

Are they friends, or are they foes?

Should AI development be stopped or pursued?

The idea of machines having their "own minds" is not without issues. It raises a number of questions including identity, ownership, intellectual property, copyright, royalties, entitlements and awards, insurance, and even financial liabilities. Take the case of an unmanned truck that causes an accident or a robot "security guard" that mistakenly lets a thief into an art museum.

There is no doubt that the application of AI would come in many shapes and forms. While AI would deliver an array of benefits, it would also introduce new challenges, which would be preposterous to completely exhaust in this book.

As it would not be possible to address all aspects of human life in view of the advances in AI, I would like to focus on some key areas in the hope of maintaining a realistic view of AI and at the same time spark insights on how we could be more prepared going forward.

May this practical guide stir further scrutiny and debate as to how we should navigate through this fateful journey that humanity and technology are taking. It is a course that clearly has the potential to completely alter the way we live in the years to come.

Welcome to the future with AI.

CHAPTER 1

ANALOG TO DIGITAL LIFE

Just a little over four decades ago, life was very different. Perhaps slower, perhaps simpler, but undoubtedly different from the way you and I live today.

Here is a snapshot of life in the 60s and 70s: We wake up to the sound of a battery-powered alarm clock. Though everyone may be rushing to get to work and school, we take time to have breakfast together. We reach for the morning paper as we sip our cup of coffee.

If we are running late for work, we pick up the handset of our rotary phone, make a call, and leave a message with the office operator.

We arrive in the office and go through a stack of paper documents. A courier goes around to pick up interoffice mail.

Our meetings are face-to-face, with attendees taking notes using pen and paper. In case of an urgent phone call, we will have to leave the meeting room.

We plan trips to the bank to withdraw cash to buy items from a grocery store on our way home.

Meanwhile, our children hang out with friends after school or spend hours in the library poring over various publications for a school project.

At home, we enjoy our favorite TV show together or relax as we listen to music from our vinyl record collection or library of cassette tapes.

Weekends find us doing chores with the kids helping out. Playtime for the children is usually in the yard or out in the street with the neighborhood kids.

We enjoy long drives and we amuse ourselves by playing games in the car or simply singing along with the radio.

Such was the analog life fortysomething years ago.

Today, our mobile phone has an alarm clock, a calendar, email, text messaging, camera, radio, and various applications (called *apps*) that provide weather and traffic updates, among many other things.

Need to go somewhere? No problem! Just use Uber[21] or Lyft[22] on our mobile phone. The app even tells us the brand of car, plate number, name and photo of the driver who will come to get us at our desired location.

Need some information? No problem!

There's Google[23] or Bing[24], popular search engines on the internet.

As for Japanese restaurants nearby, we ask Siri[25], an app on our Apple iPhone[26]. Need directions? Use Google Maps[27] or Waze[28].

Mobile phones connected to our car's stereo system via Bluetooth[29] enable us to make hands-free calls while driving.

We can get by with little to no cash for as long as we have our credit cards or debit cards. We can bank, shop, even order restaurant food from the comforts of our living room. We can send and receive text messages anytime, anywhere. We have Facebook[30], Instagram[31], and Twitter[32] apps that allow us to be in touch with a wide social network including our friends and family but also with the greater public. There are hundreds of cable TV channels to choose from. We can watch movies, videos, or listen to music using our home entertainment system, laptop, or mobile phone. The movies and video games we have today are enhanced with computer graphics, animation technology, and virtual or synthesized orchestra music.

Office communication has drastically changed with the use of electronic mail. We even have electronic calendars to help us manage our work. No need for couriers. We have Webex[33] or Zoom[34] conference calls, reducing the hassle of gathering people for face-to-face meetings.

Such is the digital life today.

For someone who grew up in the 50s, 60s, or the 70s, the changes are confounding. The digitized world has undoubtedly delivered a considerable measure of benefits to individuals: efficiency in the workplace, speed of communication, ease of commercial transactions, more accessible public transportation, and several conveniences that we now enjoy.

However, technology has its price. It has sheared the fabric of our family life. Our privacy has been slowly eroding and will continue to disintegrate. We, especially our children, are constantly bombarded with multiple and rapidly changing stimuli. Even our downtime is now filled with social media, videos, text messages, or cable TV such that our "free time" is hardly relaxing. People today are believed to be sleeping less compared to people fiftysomething years ago.

Yet throughout this exponential evolution of technology and human experience, there was a significant development that stealthily unfolded and continue to evolve before our very eyes. It was perhaps unintentional, maybe coincidental. But this monumental transformation from our analog reality to a digital life—over the last few decades—has effectively laid the foundation for what we would now call *artificial intelligence*.

CHAPTER 2

THE BIRTH OF ARTIFICIAL INTELLIGENCE

There are copious myths around the origin of artificial intelligence. Some speculate that AI came from man's innate ambition to be like gods. Whether our aspirations to create artificial intelligence are integral to our DNA or not, the reality is that talking and thinking machines in varied forms have been the subject of countless stories and sci-fi films.

Some AI historians trace back the more recent origin of AI to the 1950s when mathematicians and scientists, such as Alan Turing, Herbert Simon, Allen Newell, John McCarthy, and Marvin Minsky, began to theorize on the possibility of constructing intelligent machines.[35]

This wave of theories on the possibility of engineering AI machines in the 1950s might have been triggered by some prior technological breakthroughs: the invention of the transistor, the integrated circuit, and the first-generation computer in the late 1940s.

However valid their theories may have been, those mathematicians and scientists were evidently way ahead of their time. For in the 1950s, even several years after the transistor was born, computers were still in their infancy. Furthermore, it would require huge investment for their conjectures to become veritable.

Through the next several decades, AI would face serious doubts and cynicism, mainly because the elements to make it a reality were yet to be invented. In fact, AI historians today refer to "AI winters"[36] as seasons in the history of AI when interest and funding for AI projects dipped.

So how did we end up where we are today?

From the 1950s, it took several more years before a star was born: the first microprocessor chip. The year: 1968.[37]

Just like a celestial explosion following the birth of a new star, the 70s through the 90s witnessed a supersonic hyperdrive in microprocessor technology that continuously breached record speeds. As the dimensions of computer chips became smaller and smaller, their processing capacity grew bigger and bigger.

Case in point, the processing speed of industrial-grade, mid-range computers the size of cabinets in the 70s was overtaken by the CPU[38] speed of personal computers (PC) in the 80s. Subsequently, the speed of microprocessors doubled year on year.

The exceptional evolution in technology was not just limited to the chips. It encompassed data storage as well. From the forty-megabyte disk storage that perplexed people in the 80s, we now have disk drives that are physically smaller but have much larger storage capacity. Disks today can store gigabytes, if not terabytes, of data. And the great news is that they are available at affordable prices to businesses as well as individuals.

While this rapid technological ontogenesis was taking place, the internet grew at astounding rates. It soon became a global network of computers and devices that connected people through voice, text, electronic mail, and a suite of applications now called *social media*.

Equally important, the internet became a colossal repository of information: knowledge articles, references, journals, dictionaries, news, pictures, commercial catalogs, travel guides, videos, music, reviews, commentaries, and even the things that people talk about or buy. In other words, the internet became the global electronic library—the single biggest gargantuan storehouse of human knowledge and data!

Best of all, one could easily find information in this mammoth archive through user-friendly search engines.

This monumental explosion of data around the globe—in the internet and in millions of privately owned computer systems—became a fundamental driver in the search for ways to develop "intelligent machines." For though we have googols of data around the globe, could we build intelligent systems that could sort through immeasurable volumes of data and learn from them the way we human beings deduce patterns?

In the early days of software development, around the 70s and the 80s, a programmer would write code for the computer to wait for data (to be entered by a data entry clerk) and then write that data in a database. The actual digital data (the 0's and 1's) are written and stored on disk drives. After data is stored, a programmer would then write code to read data from the database, process the data, and produce reports on paper or display them on a terminal screen. In other words, computers just did what we told them to do.

Around the 80s, software development became more sophisticated and turned to the creation of "expert systems," where some form of intelligence was built into software programs. Expert systems were able to perform analysis, make recommendations, and even plan a set of actions and then execute the plan. Still, computers just did what we told them to do.

Nonetheless, as electronic data grew in unprecedented rates over the last two decades, academicians, scientists, business people, and technologists began to ask how these massive volumes of data could be leveraged. Certainly, no individual could sift through billions of records, establish correlations or patterns from the data set, and solve related problems. It would be ideal if computers could be "taught" how to do that.

In the old paradigm of software engineering, humans coded the rules (called *programming*) and entered the data, then the software processed the data and displayed the results. Therefore, for "learning machines" to become a reality, writing of software code had to drastically change.

Machines had to be programmed to learn like human beings—by observing, determining patterns and correlations, and then applying what was learned to solve future problems, forecast, or act given specific situations. AI technologists embarked on developing software coding frameworks and algorithms that would enable machines to derive patterns from a massive set of data and apply them to future problems. By using this framework or approach, computer systems would use statistical analysis methods to make predictions or forecasts with no explicitly written software code to perform those tasks. As a result, today, machine learning—also known as "deep learning" (having multiple layers of machine learning)—is an established subdomain of artificial intelligence.

Other AI experts took the journey of analyzing how we humans recognize the people we know in varied conditions—like the way we recognize our mothers even in a dimly lit room. The result is astonishing. AI specialists have successfully developed coding techniques that enable machines to automatically recognize people by taking specific facial features of a person from hundreds or thousands of electronic images of that person ("historical data") and use mathematical methods to compare historical data with the new electronic image of that person. This technology, called *facial recognition*, is now used to verify and authenticate the identity of individuals. Other methods to authenticate the identity of people are fingerprint recognition and voice recognition. These three methods—facial recognition, fingerprint recognition, and voice recognition—are generally known as *biometrics*, a process by which a person's identity is confirmed by using a unique physical trait of that person.

The technology used to "see" and "recognize" individuals is part of a much broader subdomain of AI called *machine perception*. The objective of this AI discipline is to equip machines with sensory capability the way we humans can hear and see. Machine perception technologies enable machines to detect obstacles, objects, or people when "autodriving" a car or an equipment. At a much finer grade and sophistication of facial recognition, there are even companies now like MPT4U that develop expression recognition software to numerically represent facial expressions of human beings—to detect anger, happiness, and even lying.[39]

Furthermore, for machines to be truly "intelligent," they must be able to interact with humans in the way humans talk to each other. AI engineers took to the drawing board and figured a way to program machines to systematically analyze large amounts of natural language data, including grammar rules. By doing so, they have conceived the natural language processing (NLP) subdomain of artificial intelligence, which involves frameworks for speech recognition, natural language interpretation (NLI), and natural language generation (NLG).[40] The old coding paradigm in natural language processing involved hard-coding the if-then rules, which proved to be too rigid. Today, NLP uses more advanced techniques like machine learning that utilize mathematical methods to make probabilistic determination on what rule to use in a situation.

Planning is one highly valued faculty of the human mind. So, for machines to be considered intelligent, they must be able to plan and subsequently execute the plan the way humans do. Planning involves determining a set of steps or procedures to achieve a goal, calculating the duration of each step, and identifying who or what is involved in each step. AI experts have developed planning algorithms by using techniques such as forward chaining inference method (also known as *working toward the goal*) or backward chaining inference method (also known as *working backward from the goal*). In doing so, the AI planning subdomain came into existence. Today, AI planning methods are used in developing unmanned vehicles and autonomous robots.

The other practical application of AI is around predictable, rules-based, and repeatable business processes. Take, for example, a process that involves reading an email, analyzing its content, and then replying to the email. Or a process that involves taking a phone call, analyzing the information gathered from the caller, and determining the appropriate response to the customer. This subdomain of AI called *Robotic Process Automation* (RPA)[41] is undoubtedly one that would be most valuable to businesses. Notwithstanding some AI technologists arguing that RPA is not AI since RPA only follows explicitly coded instructions, more advanced RPA solutions now use machine learning and natural language processing.

There are a few other interesting AI-related topics like gaming and deep neural networks that one could research in the internet.

In the final analysis, the key takeaway from this chapter is this: In just six decades—from the 1950s, when some mathematicians and scientists theorized about creating intelligent machines, to today—technology has exponentially developed such that we now have enormous amounts of electronic data, we have high-speed processors, and more importantly, we now have all the building blocks—the software frameworks and algorithms—to build machines that can hear, see, talk, learn, and think like us.

We, the human race, have finally created artificial intelligence!

CHAPTER 3

VALUE PROPOSITION

A lot of people speculate on how AI would successfully carve its way into the sphere of humanity. It is not surprising that many people are skeptical of artificial intelligence and perceive it as an enemy of humankind. There are doomsday pedagogues who are fearful of the effervescent progression of AI and its devastating effect on the livelihood of many people.

It is not my intention to confirm or allay their fears. Instead, I encourage that we all face the matter of AI head-on, gain a deeper insight on what it could deliver, contemplate on probable scenarios, and plan our practical response.

Acknowledging AI's existence is not a fatalistic approach but a realistic one. The wave that AI has created is not something that could be easily squashed. Even if it is possible to ban AI development in one country, another region would likely continue its progression.

The fact is humankind has this inextinguishable desire to satisfy its curiosity. Ethical and philosophical debates are necessary and may delay the growth of AI, but at some point in the future, AI would resurface. For as long as humans would find use for AI technologies, AI would continue to be explored.

Given my thirtysomething years of knowledge and experience in technology, allow me to share my thoughts on how AI would make its way into human life and openly discuss its potential impact. Taking into account some of my observations on current events, let me impart my thoughts on how this odyssey would unfold.

So let's begin.

Artificial intelligence, or any technology for that matter, would not survive if it could not deliver value to businesses or to consumers.

In business speak, we call that a "value proposition."

First, let's discuss the value proposition of artificial intelligence to business.

For us to be able to answer that question, we need to understand what is valuable to business.

However, for us to know what is valuable to a business, we need to consider the two key objectives of any business: 1) profit and 2) growth.

GENERATING PROFIT

Profit is the result of adding up all the company's sales for the year (called *total revenue*) and deducting all the company's expenses (also known as *total cost*). In other words, total revenue minus total cost equals net profit.

Now, to generate profit, a business has two options: 1) increase its sales or revenue, or 2) reduce its expenses or costs.

Now let's tackle the matter of increasing sales.

There are numerous ways to increase sales or revenue. You could do phone campaigns, coupon campaigns, email campaigns, social media campaigns, etc. Large companies today perform detailed analyses on the customer journey of their target market to better understand their customers' behaviors. Marketing campaigns are then tailored based on the results of their analyses.

So how could AI help companies increase their sales?

Well, with natural language processing capabilities, AI could do phone campaigns. AI could help analyze data to have a more effective coupon campaign. AI could be developed as an "Admin Aide" to take dictation and send emails to customers. More sophisticated AI algorithms could also perform market behavior analysis on social media to help you gain insights on where your target market is spending most of its time.

AI could come in many different forms. It could be an appliance, a software that you could install on your desktop or mobile phone, or software as a service available in the internet.

This is not sci-fi. Not a long shot. It's just a matter of time before these technologies become part of the normal course of doing business.

Now let's talk about cost reduction.

Any seasoned CEO, chairperson, company president, board member, or manager would tell you that in most companies, the biggest cost component is usually related to manpower (salaries, wages, and benefits). It is therefore not surprising that reduction in manpower costs is often used as justification for implementing new technologies that automate business processes or make work more efficient. These technologies would enable the completion of critical business tasks in a much shorter amount of time—with fewer people.

A sample business case is the implementation of an automated phone answering system in a global company. Such a system could result in cost savings equivalent to the salary of twenty phone operators who work on rotating shifts to cover twenty-four hours a day, 365 days a year. Such automated call answering systems are now a part of our normal course of conducting business with electric companies, gas companies, credit card companies, cellular service providers, banks, government agencies, and many other service providers.

How could AI help reduce the cost of doing business?

To answer that question—manpower costs in particular—we have to look into robotic process automation (RPA).

As RPA algorithms become more and more sophisticated, clerical functions could be taken over by machines resulting in cost reduction. Clerical functions could be anywhere from taking orders, processing service requests, and reading and responding to customer email to a little more complex function like scheduling or dispatching of any support or service personnel, calling back a customer who lodged a complaint, etc. I am not just talking of some software in the internet with some built-in process flow but an AI machine that could converse like a human being—in multiple languages.

Today, there are various AI companies, like Blue Prism[42] that provide RPA solutions to automate repetitive and rules-based tasks.

But over and above the list of benefits that companies could derive from AI, there are a few essential things to highlight. AI does not sleep. It does not take vacations. It does not get sick. It does not have any moods or behavioral issues. It could continuously operate twenty-four hours a day, 365 days a year.

So far, we have discussed how AI could help increase sales or reduce costs to generate profit. Let us move on to the other key objective of a business: growth.

BUSINESS GROWTH

Growth broadly means that the company's profit is increasing every year. Sometimes people talk of "positive growth" (increasing profit) and "negative growth" (decreasing profit). By default, we say a business is growing when its profit is increasing year on year.

Let me cite a simple illustration.

You own a grocery store and have ten workers. Last year, you made a profit of $10,000. You certainly would not want your profit for this year to be less than $10,000, would you? So to increase your profit, you either 1) plan and execute a marketing campaign by promoting your grocery store in the neighborhood to increase your revenue or sales or 2) lower your operating expenses—perhaps by reducing the number of workers from ten to nine. Then you go through the same exercise next year, to make sure that your business is growing every year.

So, business growth means that your net profit is increasing every year. It is that simple. But it is easier said than done.

To achieve growth, revenue or sales would have to grow every year—at a faster rate than the increase in costs. If revenue does not grow, meaning it's flat or decreasing, then costs would have to go down. As competition intensifies, there would be more demand for technologies that could continuously drive up sales and bring down costs. Due to increasing competition, it is inevitable that companies, both big and small, would turn to AI.

As AI continues to evolve, technology would become more affordable to companies and individuals. That said, it is possible that AI could lead to the leveling of playing field in business. This means that companies in developing countries would be able to compete with those in first world countries as the former would have access to the same AI technology, the same AI-enabled analytics, the same efficiencies, and the same decision support systems as the latter.

While it is true that large companies may have the financial edge over smaller ones, it is important to note that in business, it is not the size that matters, it's the speed. And artificial intelligence could help small companies achieve that speed of business.

In summary, what is the key takeaway here?

Any technology that businesses could use to generate profit and boost growth would likely survive the future.

As AI could clearly help businesses increase sales and deliver real cost savings to companies—in the form of manpower reduction or improved efficiencies—it is probable that AI would persist in the future.

The above discourse may sound trivial today, though going forward, the implications are far-reaching in several key areas: education, training of employees and managers, skills of workers required in the future, and ultimately the jobs that would be available. The impact of AI would be vast, not just on this generation, but also on the generations to come.

We have so far discussed the value that AI could deliver to businesses.

But what about the value of AI to consumers?

CHAPTER 4

HOMEWARD BOUND

The home is a place of rest, recreation, relaxation, and private conversations with family and other loved ones. Our home is a bastion where we feel safe and secure. It is a citadel where we can do the things that we would not do in public.

On most occasions, we want work at home to be quick and easy so that we can spend our time relaxing. This is the reason why appliances like the dishwasher, washing machine, and dryer have become so successful.

Well, the good news is that AI could make chores at home a little easier.

It is possible that one day some "home aide" device equipped with AI would be able to converse with you and follow specific instructions, like book an appointment with your favorite hair salon. It is interesting to note that Google, in one of its conferences in 2018, did a live demonstration of their AI assistant called *Google Duplex* that booked a haircut appointment.[43] The AI assistant sounded just like a person who can understand nuanced expressions.

Going back to our home aide device, this intelligent machine could also order your weekly supplies from a nearby grocery store.

On nights when you don't have time to cook for the family, it could call your favorite restaurant and order special dishes for delivery. Of course, if you prefer to dine out in a Japanese restaurant that night, the home aide could give you a few recommendations. Call the restaurant to make a reservation for you? Absolutely. The home aide could do that, too.

In case you want to cook a new dish for your family this weekend, the home aide could order all the ingredients in advance and teach you step-by-step how to cook that new dish.

Now consider this: wouldn't it be more amazing to have a robot at home that could cook for you?

In October 2016, *Forbes* (forbes.com) released an article entitled "The World's First Home Robotic Chef Can Cook over 100 Meals." Here is an excerpt from that article:

> In 2018, Moley will launch the world's first fully-automated and integrated intelligent cooking robot—a robotic kitchen that has unlimited access to chefs and their recipes worldwide. So not only can this robotic chef cook over 100 different meals for you, it will clean up after itself too![44]

Robotic hands slicing fruit.

What else could the home aide do for you?

Your air conditioning system is not working? The home aide could arrange for an aircon maintenance person to diagnose and fix the problem.

The home aide could also call the termite-control company to schedule a visit—at your preferred date and time.

As this intelligent device constantly monitors your house, it would call the police if it detects an intruder inside your property.

In case of an emergency, the home aide device could call 911 and provide your exact location to the 911 operator.

Save some money on your utility bills? Without a doubt. The AI device could help you minimize your utility bills as it has control over your lights, HVAC (heating, ventilation, and air conditioning), refrigerator, and even your sprinkler system. It could turn off the lights as well as the aircon unit when you leave for work and turn them back on ten minutes before you arrive.

Being that every home appliance would be equipped with AI in the future, each intelligent household equipment is capable of self-diagnosis. It would report any problem to the home aide, which would then arrange for service.

Watching a movie at home? Tell your AI device what type of movie you prefer—thriller, love story, documentary, sci-fi, new blockbuster—and it would make recommendations. As it has control over your entertainment system, the home aide would search for your selected movie, play it on your home theater system, and adjust the volume to your liking. It could even dim the lights for you.

In case you get a phone call while watching a movie, the home aide device is intelligent enough to figure out whether it needs to interrupt your viewing.

Just consider how this device could help people with disabilities or assist elderly people.

I am just talking of one device equipped with AI. Imagine your home with an intelligent robot—one that could wash dishes, vacuum the floor, feed the dog, and talk to you about anything that interests you. It could even play chess, poker, Chinese checkers, or a video game with you.

I am sure that for some of you, you couldn't wait to see this happen. I don't blame you.

CHAPTER 5

THE THINGS THAT MAKE US HUMAN

Beneath any AI application is an algorithm that imitates humans. Be that as it may, humans are not easy to mimic as we are a complex mix of *dramatis personae*.

Case in point, we have a long list of characteristics that differentiate us from plants and animals.

We have a level of intellect that is far superior to any other creature on earth. We have consciousness, including a deep, philosophical understanding of who we are. We have motives and intentions. We can articulate our thoughts into profound sentences and intricate diagrams.

With our level of intellect comes free will. We are not bound by some computer program or code that governs our whole being. We can choose what we want to do from the moment we wake up till the time we sleep. We can choose our daily activities, how we want to spend our summer vacation, what courses we want to take in college, what school we want to go to, who we want to marry, how many children we want, and so forth.

We have a sense of humor.

We can plan a very complicated project. We can execute the plan.

We know when we make mistakes, and we have the capacity to learn from our experiences.

We have a sense of right and wrong, good and evil. We humans have something called a *moral compass*—or conscience as some people may call it. We might not necessarily all agree on how we define right and wrong or good and evil, but the fact remains that we all understand the polarities and the gray areas. We have something deep inside—at the very core of our humanity—that tells us when something is not right, like racism or bullying. In fact, we have law enforcement organizations and a justice system that ensure we humans follow the laws, rules, and regulations that have been set by our lawmakers. And if we don't, then we get penalized.

Many believe that we humans are made up of body, soul, and spirit. Soul is said to be the seat of emotion, and the spirit is believed to be an entity that is not physical but bears our identity and consciousness. That means that our physical body may die, but our spirit will live on—perhaps in another dimension. That means that we humans understand the concept of dimensions—length, width, height, time, and other nonphysical realms.

We have emotions, such as love, hope, joy, peace, fear, anxiety, jealousy, excitement, sadness, depression, and nostalgia.

We know how to appreciate beauty and art in many forms: music, dance, sculpture, drawing, painting, poetry, literature, etc.

Regardless of type of personality, we are basically social beings. Most of us want to be around people. We want to dine with people close to us, have conversations with our friends, play sports with our teammates, hang out with our buddies, or feel the warm embrace of our loved ones.

We have families. We have parents, siblings, cousins, aunts, and uncles, and we talk of generations past and generations to come. Our family is one of the key guiding posts that tell us who we are.

We need privacy in varying degrees. We close the door when we want to be alone. We pull a friend to a corner for a private conversation. We don't want anyone snooping on our mobile phone. And more importantly, we don't like the idea of someone watching our every move.

Finally, we have life stories and experiences—happy moments, traumatic ones, tragic incidents, triumphs, defeats, love stories, travels, etc. We humans can recollect our experiences and analyze them. These are some of the elements that make up the very fabric of our humanity. These things make us distinct from any other creature in this planet.

It is essential for us to clearly understand the unique qualities that we humans possess vis-à-vis what intelligent machines could obtain. While technology would continue to advance and change our lives, there are human characteristics that machines would not be able to acquire. Perhaps mimic, but not inherently acquire.

Having said that, I have news for you.

We are now entering an era where artificial intelligence is slowly learning some of the characteristics of human beings.

The age of AI is here.

CHAPTER 8

EMOTIONAL INTELLIGENCE

We are all familiar with the broad spectrum of human emotions: love, happiness, sadness, anger, jealousy, and many others.

In the matter of emotions, let me make it clear from the very start of this chapter that machines do not have emotions. I am not even suggesting that machines would one day be able to develop human emotions the way filmmakers depict them in sci-fi movies. Even if we write the most sophisticated AI algorithms, we could only make machines sound like they have emotions. The fact remains—today and in the future—that AI is only made up of electronic components and some software code.

That said, machines do not and could not have motives in the way that humans have intent. Again, we could craft the most advanced AI algorithms and make machines sound like they have motives, but the reality is AI is just some bits and bytes running on computer hardware.

While we humans have life stories and experiences that influence the way we feel, think, and behave, machines do not have those kinds of stories to tell. Machines only have a set of instructions to follow. That is their world.

Nonetheless, it is interesting to note that AI technologists have started development in the area of sentiment analysis.[45] That means that computers (running some software code) are now able to distinguish "positive" from "negative" comments or text. Sentiments like angry or happy could now be detected by machines by analyzing text. They could also detect the degree of the sentiment (e.g., angry or very angry).

Some examples of sentiment analysis platforms include IBM Watson Tone Analyzer API[46], Bitext Text and Sentiment Analysis API[47], Emotion Analysis API by Twinword[48], and Brandwatch[49].

Output from sentiment analysis could be consumed by marketing specialists to determine the effectiveness of their marketing campaigns based on comments on social media.

Insurance companies could calculate the risk associated with insuring a piece of property by performing sentiment analysis on social media posts from people in the neighborhood.

Now consider that in the context of natural language processing capability of AI, it is likely that you would one day have conversations with a machine that would ask, "Why do you seem angry?" or "Why are you sad?" Such questions from an intelligent machine—performing real-time sentiment analysis—would be part of an ordinary conversation with a human being in the future.

Let us pause for a moment and analyze this scenario.

It is said that the deepest level of human conversation or communication is when we reach a point where we talk about our feelings. We humans get by, day in and day out, with casual conversations about the weather, news, sports, and other topics of interest. But at certain points in our life, we need that depth of conversation whereby we could freely talk about how we feel—without being judged.

I know it would really sound absurd to some that we are even thinking of a person talking to a machine about his or her feelings. Yet this is not a remote possibility. Today we have commercially available devices or appliances that could converse with human beings.

Furthermore, some robotics companies, like Hanson Robotics, are now developing "social robots" designed to be companions to humans.[50]

With ever-increasing stress in daily living, people may resort to robots for companionship, much like some prefer to live with their dog or cat.

If you think about it, machines don't judge, don't lie, and don't get cranky. They require minimal maintenance.

Having a talking machine could even be entertaining and mind stimulating as robots could discuss most any topic under the sun. The elderly as well as people with disabilities who live alone could benefit from interactive machines. People who are bed-ridden for long periods could find comfort even in a simple electronic appliance or machine that could read to them stories or perhaps talk about the latest news. Due to the aging global population, robotics companies believe that the next biggest thing is elder care robots.

Think of robot caregivers who could be available 24/7. Robots could help with problem-solving and they could even call for emergency services in a crisis situation. The possibilities are endless.

Finally, people of any age going through personal crises could find relief in having a chat with "someone." While some may argue that machines do not have and will never have the level of emotional intelligence of humans, this cold mechanical machine we call "robot" may just be the very thing that could save the lives of many people in the years to come.

CHAPTER 7

THE ART OF MAKING ART

Art—in all its forms—has always been attributed to humans' desire for expression. It has been widely accepted that humans have souls—the seat of human emotion.

Take, for example, music. Johann Sebastian Bach was believed to have said, "All music should have no other end and aim than the glory of God and the soul's refreshment."[51]

Music educators would argue that music must be intentional. And being that machines and animals do not have intentions or motives like human beings, music could only come from human beings—and only on occasions when a person intentionally does it.

That means that while footsteps might create some rhythm, they do not automatically qualify as music—unless intentionally done to create music. Similarly, snoring could have some cadence, but it would hardly be considered music as the person creating the sound is not conscious to make it intentional.

That is also the reason that while chirping of birds might sound pleasant to one's ears, it doesn't qualify as music. Why? Because birds, or any animal for that matter, do not have intent the way we humans have. As such, music authorship could only be attributed to humans, not to birds, dogs, cats, or lions. Therefore, copyright[52] could only be granted to humans. Royalties[53] are only distributed to people. And music awards are only given to human beings.

AI IN MUSIC

In the last three decades, there were three parallel developments that laid the groundwork for the application of AI in music.

The first is the digitization of music. Music, for a very long time, has been known to be analog—in the form of sound waves. Music was recorded on tapes. The breakthrough happened in 1982 when Sony and Philips developed a technology that could store sound or music as bits and bytes on optical compact discs (CDs). It was a topic of conversations and speculations in music circles around the globe. There were valid reasons for the buzz. The digitalization of music meant that songs could be copied from one CD to another and there would be no loss in sound quality—unlike its predecessor, the tape. Digital music also meant that songs could be sent to anyone anywhere in world through electronic mail. That would translate to untracked use of music, copyright violations, and lost revenue for music creators.

The second development is the advancement in digital sound recording. Soon after sound was successfully digitized, audio recording manufacturers created sound recording equipment that could record music in digital format. Over the next few years, recording studios across the globe started offering digital recording to their customers. This wave led to the creation of massive amounts of digital music, digital samples, and digital sound libraries.

The third development is the rapid progress in the area of synthesizers and music arranging software. The first-generation synthesizers, also known as *synths*, developed by Moog and Hammond manipulated sound waves to create synthesized sounds. The digitization of music, the birth of digital sound recording, and the accelerated growth of digital samples and sound libraries, made the synthesizer industry a happy beneficiary. As a result, synthesizers today sound like real instruments—real pianos, real drums, real trumpets, real string sections, etc. In fact, the synthesizer technology has become so advanced that there are synths now that have preset rhythmic patterns. Synths can now perform arpeggios and have guitar-strumming algorithms.

Moreover, due to the enormous repository of digital music, digital samples, and digital sound libraries, artificial intelligence now has a lot of data to learn from. In fact, today there are numerous songwriters who work with AI software to create music.

Is it possible that AI would one day create music on its own? It already happened. Futurism.com released an article in August 2017 entitled "The World's First Album Composed and Produced by an AI Has Been Unveiled."[54] However, AI-generated music would create a new host of challenges, questions, and debates in the music industry. New rules, regulations, and laws must be created around AI-related authorship, copyright, royalties, and even music awards.

But the application of AI doesn't stop there.

Robot playing on the piano.

AI IN FINE ARTS AND LITERATURE

AI technologists are now exploring the use of AI in fine arts. That means feeding AI with hundreds or thousands of paintings (i.e., digital photos) to learn from and guiding AI in learning the fundamentals of visual composition. Is it possible that AI would one day be able to create a painting given a theme? I would say it is not just possible. It already happened.

On October 25, 2018, the world-famous Christie's Auction House in Manhattan sold an AI-generated artwork entitled "Portrait of Edmond Belamy," created by Generative Adversarial Network, for a whopping $432,500, beating original estimates of between $7,000 and $10,000. The AI-generated painting was apparently created by an AI software developed by three French technologists. The painting is surrounded by controversy that part of code was "borrowed" from a nineteen-year-old high school graduate.[55]

This successful sale is clearly a breakthrough in AI-generated art. Its value could be attributed to its being one of the first AI-generated artworks. And now that there is a controversy surrounding it, there is a story to tell about the painting.

Other AI experts are now looking at applying AI to literature. This involves giving AI many literary works to learn from and guiding AI to learn the fundamentals in writing a poem, for example. Is it possible that AI would one day create short stories or write a poem? I would say it is possible.

WHERE THE REAL VALUE LIES

Now let's ponder the essence of art.
 What really makes art an art?
 Where does the value of an artwork lie?
 Is the real value of the artwork in the artwork itself? Or is the value of an artwork found in the creator of the artwork?

I know this may sound philosophical, but please bear with me for a moment.

Beethoven's Symphony No. 9 is believed to have been completed by the master composer when he was already deaf. It was said that during the premiere of his famous symphony, Ludwig had to be turned around to the audience that was thunderously applauding at his work—proof that his hearing was totally gone.[56]

Chopin's poignant and heart-wrenching piano concertos were written by the young composer between episodes of illness that were later attributed to possibly tuberculosis. The young Frédéric François Chopin left his beloved Poland at the tender age of twenty and traveled to Paris. He then got involved with a scandalous woman and famous novelist, Amantine Lucile Aurore Dupin, who was also known as George Sand. Chopin eventually died at the young age of thirty-nine in 1849 with a much-celebrated and star-studded funeral rite in Paris. It was said that Chopin, just before he died, asked his sister to physically bring his heart back to Poland, where it is now buried in the grounds of the church of the Holy Cross in Warsaw.[57]

The Impressionist movement in the nineteenth century that was started—perhaps unknowingly—by young artists, such as Monet, Renoir, Manet, Sisley, Pissarro, Degas, and Cézanne, was a result of repeated failures to be accepted by the high-brow Paris Salon that judged artwork according to the accepted standards during that time. Their paintings were believed to be "unfinished" or lacking in the basic elements of a masterful artwork.[58]

The famous painter Pablo Picasso was known to have had a Blue Period, roughly from 1900 to 1904, during which he painted monochromatic paintings in various shades of blue—attributed to the financial hardship he experienced during those years.[59]

The Renaissance masterpiece called *David*, which was sculpted by the Italian artist Michelangelo di Lodovico Buonarroti Simoni—more widely known as Michelangelo, was done in utmost secrecy by the master artist, who was believed to have studied the details of the human anatomy by working with cadavers.[60]

What can we draw from the above?

The value of a piece of art—whether it is music, painting, sculpture, or other forms—lies in the stories behind the artwork. I am talking of people stories, life stories, failures, eccentricities, anecdotes, and the situation the artist faced at the time that he or she created the masterpiece.

In fact, art museums, galleries, auction houses, collectors, curators, and other art enthusiasts diligently research the history of an art piece as it is an essential part in assessing the value of an artwork. Why? Because it is the human drama, the real-life story behind an artwork, that underpins its true value.

A song, a painting, or even a poem produced by a machine may be a remarkable technological breakthrough, but there is very little or no story to tell about its creator and the circumstances around the artwork.

In my humble opinion, artwork produced by humans will continue to hold value to humans.

The bigger question is this: "Will AI-generated artwork garner the same level of interest from humans and demand the same value as that of art produced by humans?

The October 2018 sale of an AI-generated artwork in the world-famous Christie's Auction House in New York City is undoubtedly a sign that it could, though it is too early to tell.

We must remember that what drives the market value of an item is strong demand coupled with limited supply. In the case of AI-generated artwork, the AI code has the potential to generate artwork much faster than a human being. Ergo, supply, in this case, could hardly be considered as limited.

While AI is generating a lot of interest today, even in the art circles, it would be very difficult to say without a doubt that the interest would be sustained for decades to come. It is also a challenge to conclusively state that the value of AI-generated artwork would continue to rise in the coming years.

Whatever happens to the value of AI-generated artwork going forward, there is one thing that I could state with strong conviction: human artwork will continue to be valuable to humankind.

We may one day find a spot in top museums for AI-generated art (some already have), but the work of the masters—including those yet to be recognized—will always hold a place in the magnificent halls of art museums around the world. After all, art reflects the human heart and soul and, therefore, represents who we are as a people.

Let me pause here.

Think about where the world is going.

How then should you gear up yourself and your children for that journey to the future?

CHAPTER 8

NOW THAT'S ENTERTAINMENT!

We humans love to be entertained. Some of us enjoy entertaining others through music, dance, or drama. It is part of our human nature.

The word *entertainment* has taken a whole new meaning today.

Entertainment now covers television, cable, video subscription services, film, animation, music, theater, video games, and virtual reality.

As AI involves ingesting lots of data, establishing patterns, then applying the "learning" to future problems, how could AI be applied to the entertainment industry?

To answer this question, we need to understand the process in the entertainment industry.

CREATING AND MARKETING CONTENT

First, an entertainment company will conduct a marketing study to determine what viewers, listeners (in the case of music), or gamers want.

Second, a theme is conceived, developed, and budget is approved. Production is planned and executed. Then, postproduction is undertaken.

Third, marketing of the product is kicked off and the final product is released.

Let's discuss the first phase: marketing study.

How would an entertainment company know what film genre, type of music, or game is selling?

Marketing data can be obtained from a variety of sources today. In the case of movies, there are sales reports from movie houses, viewing information from video subscription services, click rates and likes from social media sites (say, on movie trailers), DVD sales, and comments on social media posts. In the case of music, there are music download sales information, music subscription stats, playlists, radio airplay, among many others. For games, you have store sales stats, data on game downloads, and many other sources.

The challenge is correlating the above data to the profile of the viewer, listener, or gamer. What this means is that while entertainment companies could get, for example, the total sales from movie houses or click rates from social media sites, there is no single source of information on age, gender, ethnicity, profession, and other details on each viewer, listener, or gamer.

Any marketing student would tell you there is a need for that detailed level of information. In the field of marketing, there are concepts called "target market segment" and "market segmentation." A market segment is simply a band of people who share some common characteristics (interests, age group, activities, etc.) grouped together for marketing purposes.

Today, that information is almost impossible to gather. There are numerous restrictions and privacy concerns that people would raise against the harvesting of such personal information.

Nonetheless, AI can be used to perform analytics on whatever data is available to draw correlations and patterns. The results could be used to understand market trends to help plan for future releases of films, music, or games.

Now let's talk about the second phase: production and postproduction.

It is possible that the creation of stories and plots for film and games in the future would be aided by AI. As for music, AI is already being used by some songwriters in producing new materials.

In the entertainment industry, production and postproduction processes require a lot of planning and coordination. In the case of filmmaking, for example, there's scheduling of equipment, booking locations, arranging logistics for talents and crew, ordering supplies, booking caterers, etc. An AI-enabled "production aide" device could assist production managers in coordinating all those activities.

As far as technologies used in film and video game production, movies and games in the future would incorporate AI, 3D, 4D, and virtual reality (VR) technologies. I believe VR technology could be a fertile ground for AI.

Moving forward, let's discuss the third phase: advertising and release of the final product.

As for advertising, it is important to note that there are companies today that specialize in the application of data science on advertising. Artificial intelligence could further enhance data science to a level of pinpoint accuracy.

As for the release of the final product, AI could be used to coordinate logistics and distribution, analyze sales, and provide valuable feedback on the effectivity of advertising initiatives.

THE ULTIMATE VIRTUAL EXPERIENCE

Now that we have discussed how AI could be used by the entertainment industry, let's see what the viewer, listener, or gamer experience would be like in the future—10 to 20 years from now.

First, I believe that you would still have a choice of watching movies, listening to music, or playing games using your home theater system, laptop, or mobile phone—with virtual reality (VR) headsets. The only difference is that the VR headset a few decades from now would be much different. Furthermore, new devices other than TV, laptops, and mobile phones, could surface in the next several decades.

The VR headset of the future—or whatever name they would call it then—would be able to give you an entertainment experience that would involve more senses: touch, sight, and sound. I am not certain about smell and taste. Anyhow, you would feel the sensation and the action—the sword fights, horse rides, or the supersonic flight as you soar through the skies.

The movies and video games of the future would be highly interactive. This not only means that you could be part of the movie, but you would also be able to interact with others who are watching the same movie or playing the same video game. While one would argue that this kind of technology is already here, the biggest difference is the use of AI in movies and video games. For example, AI would be able to dynamically alter the story, change the landscape of the game, or introduce new characters, plots, or objects in the movie or video game.

With the incorporation of AI in video games, AI could generate a completely new dimension or parallel universe for gamers. The games would not be hard coded. AI would learn from a massive repository of data from true reality and randomly create characters, situations, landscapes, objects, animals, and even transportation in this parallel world.

As for music listening experience in the future, you would be able to watch your favorite band's past concerts or your choice orchestra's past performances using a VR headset and get the "live" concert experience—just like you are there in one of the theater seats. Your virtual reality headset would be equipped with a sophisticated speaker system that would deliver the same ambience as being inside the concert venue. Not only that. The future VR technology could even give you the ultimate virtual experience of being the conductor of your favorite orchestra.

Now that's entertainment!

CHAPTER

MINDING MY OWN BUSINESS

There is something in human nature that demands varying degrees of privacy, depending on culture, personality, and other factors.

We don't like people watching us while we eat. We don't want strangers listening to our conversation with a friend. We don't like the idea of other people going through our Inbox.

While there is much talk about the millennials slowly learning to let go of their privacy, they would still abhor parents who snoop on their social media accounts or ambush their mobile phones.

Privacy, therefore, is innate in human nature, regardless of age, race, education, or creed. No matter how people try to downplay its importance, there is no such thing as a totally transparent "public" person. Generally speaking, there is filtering happening before we speak. Sure, we sometimes say, "Oops, slip of the tongue." But even so, there are many thoughts throughout the day that we just keep to ourselves. That is a way of keeping our privacy.

THE MATTER OF PRIVACY

Just as we talk about us humans needing some privacy, the truth is that technology is shaping an environment that is sneaking its way into our private world.

Social media has undoubtedly enabled sharing of information, thoughts, moods, comments, and even photos and videos that were once considered personal.

"What's on your mind?" is a familiar line in one global social media app.

People can easily determine your political, social, ethical, and religious views from your posts and comments. And it is not news that your whereabouts can be tracked through your social media posts.

But we are just seeing the tip of the iceberg. Underneath, there are companies that provide digital marketing services—like displaying ads wherever you go in the internet. That simply means following you through all your digital activities—social media, email, website visits, internet searches, etc.—and displaying ads wherever you go. There is, in fact, an official term in digital marketing for that. It is called "retargeting" or "remarketing."

The idea is very simple. You visit a website of a famous apparel store. You click on, say, winter coats. After spending a few minutes on that website, you decide to read your email. You then find a new email in your Inbox from the same apparel store reminding you to do your online shopping now while it has an ongoing sale on the items you happened to browse just a few minutes ago. You wonder how that apparel company knows about your email address. *Perhaps it was just a coincidence,* you think.

Your suspicions grow when you go to your social media account and find several ads from the same apparel store showing in your newsfeed every now and then. You suddenly realize that your favorite apparel store must know your email address and your social media accounts. You wonder what else it could know about you.

The fact is there are groups that harvest information on users in the internet and sell the information to companies that the latter could use for marketing campaigns.

It is almost inevitable for anyone, in this day and age, to have some form of electronic record or profile in some system or somewhere in the internet. When you create an email account, for example, you are asked to supply your phone number. When you buy a mobile phone, the phone company asks for your home address and a few other personal details.

When you set up autopay for your electric bills, you are asked to provide your credit card or bank account details. When you apply for a credit card, you are asked for personal details like annual salary and job title.

As we move into the future, you could expect that more and more business and personal transactions will require information to be stored in digital format. This could mean your financial information, educational background, work experiences, passport information, finger scan information, medical history, and yes, even your DNA information.

Some people might argue that capturing DNA information is highly debatable as it is a violation of privacy. Nonetheless, the lines of privacy are slowly getting blurry. On February 6, 2019, the *New York Times* released an article stating that FamilyTreeDNA, an at-home DNA testing company, "apologized for failing to disclose it was sharing genetic information with the FBI to help solve rapes and murders."[61]

It is possible that we will see the day when the DNA profile of every newborn child will be stored in some tiny chip.

A chip embedded in your skin? Possibly. But this would not easily happen as you would expect people to push back. But there is one major issue, which I will explain later in this chapter, that would arise and subsequently drive governments to implement this type of personal identification.

EQUAL AND OPPOSITE REACTION

It would not be too difficult to imagine a future where transactions are completely electronic, meaning, there is no need for cash. Even today, we have most, if not all, of our financial records in various computer systems. We hardly need to go to a bank to withdraw or deposit hard cash. We pay our restaurant bills, groceries, office supplies, utility bills, and phone bills with our debit or credit cards. Even cryptocurrency is now recognized in many countries.

That said, while personal records in digital format have proven benefits, they also have risks. There are people out there who would try to steal your digital information, spoof your identity, request a new debit or credit card, submit a loan application, or purchase items—at your expense.

Such is the reality today that there are companies specializing in the protection of your identity and credit card information. Fraud alerts are sent to you in the event of a suspected breach of your security.

The good news is that in the future, it is possible that AI will be used to secure your personal information from hackers and criminal elements. AI could also be trained to continuously scan for possible security breaches and immediately address the risks to your personal and financial records. It is like having an IT security expert guarding your personal information round-the-clock.

There are companies today, like BioCatch, that use AI to perform "Continuous Authentication" to ensure that the user who has passed the user login screen is indeed the rightful owner of that account.

How?

Here is BioCatch's secret recipe:

> The solution selects 20 unique features from its 2,000+ behavioral profiling metrics to authenticate a user — without any disruption in the user's experience. The features are selected according to highly-advanced machine learning algorithms, which are employed to maximize the profiling process. After a few minutes of user activity, a robust user profile is built. Once established, the system can detect anomalies and suspicious behavior at an extremely high-level of accuracy and low rate of false positives.[62]

That's all good news. What's the bad news?

It is said that for every action, there is an equal and opposite reaction.

It is possible that tech-savvy hackers would also use AI to hack systems and steal personal information of other people. Imagine the catastrophic effects of well-crafted AI algorithms that could learn from hackers' cookbooks and stealthily penetrate systems with no human intervention. That's a completely different level of sophistication.

This possibility is not remote. We already have computer viruses today called *ransomware* that could lock down your data files and documents so that you can't access them until you pay ransom money to the hackers.

The application of AI in cyberspace would bring information security to new heights and depths never seen before.

You can foresee the complexity that such a scenario would introduce in terms of legislations, jurisdictions, law enforcement, and criminal justice, not to mention the amount of technology that would be involved in tracing back from the "crime scene" to the criminal.

It is critical to remind ourselves that the culprit is not AI but the people who develop the AI algorithms to commit the crimes.

What am I saying here?

As the world moves toward a fully digital era—where everything will be done electronically and every electronic device will be interconnected to a huge global network of devices—there could be a new wave of highly sophisticated criminal activities using AI in cyberspace.

In reaction to these cybercrimes, it is likely that governments, companies, and other groups would use AI and other technologies to protect their constituents from unscrupulous individuals.

BODY CHIP

To control the proliferation of cybercrimes around the globe, governments would be forced to implement drastic measures to track and preempt cybercriminal activities. One possible way to do that is through a global or national ID system using a device that could not be easily tampered or lost. That ID system would have to be something small, could contain a substantial amount of personal data, and be embedded into a part of the human body. For lack of a better term, let us call it "body chip."

The body chip would be able to draw power from multiple sources, including sunlight and your body heat. It would fail to work if it is removed from your body. Any transaction that you make would be authenticated by multiple factors, including a code from the body chip, finger scanning, facial recognition, and even voice recognition.

This body chip would be able to receive and transmit data, meaning it could interact with other electronic devices. It would serve as your wallet, your driver's license, your office ID, your social security ID, your medical history folder, your passport, and your debit and credit cards.

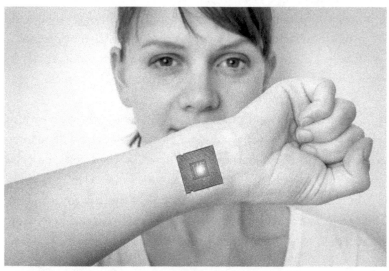

Bionic chip (processor) implant in female human body.

Your body chip would interact with your laptop and automatically log you in, given some voice commands or even by just a wave of your hand. Your body chip would be used to authenticate you when you buy an item on the internet. It would interact with a credit card machine to pay your bill in a restaurant. In other words, the body chip would yield a lot of benefits in terms of efficiency and security. However, this could also mean that you would be completely traceable in the digital world. I know this sounds like stuff in sci-fi movies, but this technology is already here.

On April 3, 2017, the *Los Angeles Times* published on its website an article entitled "Companies Start Implanting Microchips in Workers' Bodies."[63] Here is an excerpt from that article:

> What could pass for a dystopian vision of the workplace is almost routine at the Swedish start-up hub Epicenter. The company offers to implant its workers and start-up members with microchips the size of grains of rice that function as swipe cards: to open doors, operate printers or buy smoothies with a wave of the hand.
>
> "The biggest benefit, I think, is convenience," said Patrick Mesterton, co-founder and chief executive of Epicenter. As a demonstration, he unlocks a door merely by waving near it. "It basically replaces a lot of things you have, other communication devices, whether it be credit cards or keys."
>
> The technology itself is not new: Such chips are used as virtual collar plates for pets, and companies use them to track deliveries. But never before has the technology been used to tag employees on a broad scale. Epicenter and a handful of other companies are the first to make chip implants broadly available.

Mounting efforts to secure cyberspace, protect individuals, companies, and electronic assets, as well as enhance law enforcement will clearly have an impact on privacy. In fact, there is truth in scuttlebutts that there are some countries today that are using real-time facial recognition to track the identities of individuals walking around the monitored area.

In attempts to secure electronic assets, even the latest mobile phones will now authenticate its owners through facial recognition.

What's the benefit? It will render the phone useless to a thief. On the other hand, this also means that through your phone, people can now determine exactly "who you are, where you are".

What if you leave your mobile phone at home?

Satellite technology has now made it possible to track vehicles almost anywhere on earth. And just by adding another feature in your car that could recognize your face or your voice, your whereabouts could undoubtedly be followed.

These technologies sound fascinating, but have you ever considered their impact on people? What about the psyche of a person who is aware that he is constantly being monitored? Have you thought of what people today, moreso in the future, must do to enjoy a "me-time" or some private moments with their loved ones?

While there is pushback against the use of AI and other surveillance technologies to track people today, we could be just a few decades away from seeing privacy becoming a thing of the past.

If that assumption comes to fruition, then would that future be a rosy picture of a free world that we envision for our children and the generations to come?

CHAPTER 10

WHAT ABOUT MY JOB?

There are a lot of myths around the term *artificial intelligence*. Mainly attributable to science-fiction movies, a layperson today is likely to think that AI is that gigantic, intelligent machine that would one day rule the world and all humans would be working for this electronic and heartless master.

It is essential that we come to a clear understanding of what AI is all about, peel back the mystery surrounding it, and analyze what it really is at its core. Then and only then can we determine how we humans should prepare ourselves and the generation to come, in view of what is inevitable.

Let's first talk about the positive side of AI.

Artificial intelligence would generate a new set of jobs that would, of course, require a new set of skills.

Before we proceed, it is necessary to note that the generation of AI algorithms today is called *Narrow AI*. That means that the AI we have today is not a superintelligent machine that could figure out anything and everything in this universe.

In other words, each of these AI algorithms would have to be designed for a specific purpose, developed, tested, maintained, etc. This means that there would be jobs around AI software design or architecture, AI software development, AI code testing, AI project management, AI maintenance and support, etc. There will be a need for people who can develop and execute roadmaps for AI products and services, conduct market research, plan and implement marketing campaigns, and of course, perform all backend support functions—HR, accounting, finance, IT, etc.

IT security would be another area that would grow with the AI industry.

If you add robotics to the picture, it would not be a surprise that the robotics industry would likewise spur growth in job opportunities around robotics strategies, sales and marketing, physical design, testing, product quality assurance, installation, and of course, post-installation maintenance and support.

Furthermore, as I stated in previous chapters, AI needs data to ingest. What does it do with the data? It "learns" or derives patterns and correlations from data that it is given and uses that learning to solve future problems. For AI to be more effective, consistency of data is imperative. But, inconsistency of data is almost inherent to human nature—typographical errors or simply inadvertent variations in data entry. This means that the quality of data used by AI needs to be managed. This translates to jobs around data management: data preparation, data quality assurance, data transformation, and data analysis.

For AI projects to succeed, it is critical for these undertakings to have business analysts, project managers, and program managers specializing in the field of AI.

It is not a remote possibility that information technology departments across major industries would soon have AI working groups that would evolve into a more formal AI section or AI unit to be headed by a vice president for AI or a director for AI.

Demand for these types of skills would exponentially rise so that colleges, universities, and even training centers would offer AI degree programs and certificate courses.

Obviously, a new industry around consulting and contracting AI implementations would rise. Some companies could also offer "AI as a service" (AIaaS) in the internet to companies that do not want to create their own internal AI service.

These AI consulting companies would have industry vertical specializations—health care, utilities, banking and finance, travel, hospitality, logistics, military, national security, home application, office automation, education, etc.

Again, it is crucial to note that the AI we have today is far from what is called *General AI,* meaning, AI machines that possess human intelligence. So far, there is no single AI solution that will satisfy a broad range of AI requirements. As such, AI would open new business and job opportunities across a wide range of industries.

In the health care industry, AI could be used in patient registration, patient scheduling, billing, etc. AI could potentially accelerate research. AI could assist in various patient care functions. AI could even be used for high-precision surgeries, digital imaging analysis, and medical training, among many others.

In the finance industry, AI could be applied to data science, fraud detection, anti-money laundering, loan processing, and many others.

In the insurance industry, AI could be applied to insurance application processing, claims processing, and fraud detection.

In the air travel industry, AI could be used to optimize revenue by analyzing flight routes, determine the most cost-effective flight schedules, evaluate marketing campaigns, etc.

In the hospitality industry, AI could be applied to hotel booking, market analysis, managing timely delivery of supplies, arranging tours or concert tickets for hotel customers, billing, and many others.

In the logistics industry, AI could be applied to equipment planning, routing optimization, manpower scheduling, booking, ticketing of customer-facing activities, billing, and many others.

In the area of security, the application of AI could deliver benefits to the military, national or homeland security, the FBI, the CIA, and various state and city law enforcement agencies. These sectors would need experts on artificial intelligence for surveillance, countersurveillance, and even AI-related forensics.

As for the application of AI to automate common house chores, we discussed that in chapter 4.

In terms of applying AI in the corporate world, AI could be developed to assist executives in maintaining their calendars, taking dictation, sending emails, and booking flights, restaurants, hotels, car rentals, etc. And more importantly, it could perform data analytics that would support management decisions.

AI could also be used in automating IT functions—systems monitoring, automatic restart of servers or processes running on servers, intrusion detection, and many others.

You can clearly see in the above that there is a very wide range of industries where AI could be applied. Each industry will have its own set of requirements. This means that there will be areas of specialization within the field of artificial intelligence. That translates to many job opportunities around AI.

That is the good news.

What about the not-so-good news?

As you can see, the best use of AI is in the area of clerical functions where activities are logic-based, rules-driven, repetitive, and predictable.

So, think about your work today.

Does your work involve routine activities?

Are the activities predictable, and could they be outlined in a set of procedures?

Could each step in that procedure be defined in an if-then rule? For example: "If the email complaint is about refrigerators, then forward the email to the Refrigerator Maintenance email group."

One could argue that a lot of jobs involve activities that are logic-based, rules-driven, and repetitive; therefore, all jobs will be replaced by AI. Not necessarily true.

What will determine the need for a human being to perform a

specific job is this: the criticality of outlying incidents that may arise outside of what was defined as "routine."

A good example of this is the job of an IT security specialist. Sure, you could apply AI to routine activities relating to computer security, such as security patching and analysis of security alerts. But the moment you have a major security incident, you need a human to manage the troubleshooting process—even if you have sophisticated AI tools—especially if the security breach affects your most critical business systems and data. A security breach could seriously damage the reputation of your company; therefore, you would need an intelligent human being to work with AI to address the situation.

Talking about the negative impact of AI on blue-collar workers, how about the jobs of drivers?

Automotive manufacturers are now designing and producing cars with autosteering features. While the technology may be in its early phases, you could expect that it would continue to mature over time. One would think that it is possible to one day see cars, trucks, buses, trains, and perhaps planes that are so "intelligent" that they would be completely on autopilot.

Self-driving car.

An "intelligent car" could also mean that your vehicle could call 911 or drive you straight to an emergency room in case you suffer a heart attack or encounter any health-related emergency inside your car. This autopilot feature would be extremely helpful for the elderly and people with ailments and disabilities.

Considering this autosteering technology, does this mean that taxi drivers, bus drivers, truck drivers, and other professional drivers would one day be out of job? In my opinion, public transportation would continue to have human drivers—until such a time that autonomous vehicles gain the complete confidence of humans.

Why?

Owners of taxicabs, trucks, and buses would still want human beings to drive their cars to avoid potential liabilities, though one could argue that the risks could be mitigated by an insurance policy. Be that as it may, insurance policies can only cover financial liabilities, not criminal liabilities.

Moreover, in case their vehicles break down on a highway or somewhere in the city, their human drivers can troubleshoot or call for road assistance. The same is true for trucking companies or large companies that have their own fleet of trucks.

Going forward, as AI technology matures, we would see more and more unmanned vehicles. When AI reaches that level of maturity, the job prospects for human drivers could dramatically change.

And what about the airline industry? Well, the newer planes now have autopilot capabilities. But then considering the cost of a plane and the responsibility around flying hundreds of human passengers, I don't think that airline companies would take the risk of running their planes on complete autopilot mode (i.e., with no pilot in the plane). Besides, I don't think passengers would be too excited to ride a plane that has no human pilots. Again, AI technology would have to gain the full trust of humans before we see the day when commercial planes would fly without pilots.

Let's pause at this point.

Give yourself time to ponder the possible impact of AI on your job.

Contemplate how you should guide your children on their career choices.

CHAPTER 11

TEACH YOUR CHILDREN

Education is one major production in the life of every person. In the United States, one spends six years in elementary school, three years in middle school, four years in high school, and then another four years in college. That's a total of seventeen years leading to a bachelor's degree.

If one decides to pursue post-graduate education, it means an additional year or two for a master's degree and even more years for a PhD or a medical degree.

In the US alone, there are about 89,000 primary education (elementary) schools, around 99,000 middle schools, and about 37,000 high schools.[64] That said, there are millions of teachers in the country.

The norm today is a student goes to school, attends face-to-face instructor-led classes, and takes quizzes or exams along with all of his or her classmates. A student would have to schedule a meeting with a teacher in case there is a need for guidance on a lesson or an assignment.

PERSONALIZED LEARNING

There have been talks about adjusting the pace and level of complexity of lessons, depending on the student's speed of comprehension and other factors. But it requires a lot of work for the teacher. It means that there would be special classes for slow learners, another set of classes for the average learners, and another set for fast learners. That also means several sets of quizzes, exams, reaction papers, etc.

In most high schools in the US, advanced placement (AP) classes are offered to fast learners. These AP classes use college-level materials, curricula, and examinations. The high school AP classes are, in fact, recognized by colleges all over the US and are accepted as college credits.

Still, a lot could be improved to tailor lessons to each student's speed and level of comprehension. It is almost impossible for one teacher to do that for all twenty or thirty students.

Well, AI could help in this area. But how?

Most universities offer online certificate programs that you could take at your own pace. The course materials may be a combination of videos, reading materials, and course assignments. At the end of each course in the program—that is after you go through all the videos and reading materials—you do your assignment and submit it to your instructor online. The instructor checks your assignment and grades your work. Should you have questions, you could chat with the instructor at times predefined by the teacher.

It is possible that this model for teaching a course (i.e., online) could be used at all levels so that it could be the norm from elementary all the way to college.

Does it mean that, in the future, students would no longer need to go to school? Not exactly. What I am saying is that families would have the option to physically send their children to school, or do self-paced online training from home, or do a combination. They would have a choice. That means that children with physical disabilities would be able to get education even if they stay at home. The value proposition is not just the flexibility in location but, more importantly, the ability to pace the lessons based on the student's level and speed of comprehension.

So how could this be achieved with AI?

In the future, it is possible that "teaching aide" devices equipped with AI could be provided to students. These devices could guide a student through the course materials (video, reading materials, etc.) just like the way a human teacher would teach a student. The device could discuss the lessons with the student, show the lesson material (video or text), and then ask the student a set of questions. If the teaching aide device determines that the student needs a review, it could play back the lesson material. If the student requests to talk to a human instructor to get some clarification, the teaching aide would connect the student to a designated human instructor.

Now envisage all that with virtual reality headsets. The whole teacher/student experience becomes almost real.

On November 1, 2018, CNN (www.cnn.com) released an article on its website entitled "Can Virtual Reality Revolutionize Education?" The article included an interview with Guido Kovalskys, chief executive and cofounder of a US-based education technology (edtech) company called *Nearpod*. Here is a quote from that article:

> According to Nearpod's figures, more than 6 million students in the US and beyond have experienced its VR-based lessons, such as virtual field trips, after it began offering the service two years ago.[65]

Furthermore, imagine the benefits of using AI and VR technologies to train medical students, surgery interns, science majors, or students in various engineering fields, for example.

We could expect this trend to continue to rise in the coming years until it becomes a regular element in the educational system.

But that is all on the technology side.

Teenage students with virtual reality headsets in a classroom.

REFOCUSING EDUCATION

I believe what is more essential to discuss is how education should look like in the future, given advancements in technology, especially around artificial intelligence.

You would have gathered from previous chapters that AI has the potential to take over jobs that are clerical, predictable, rules-driven and repetitive in nature. Some may call them blue-collar jobs. That means clerks, cab, bus, and train drivers—as driverless vehicles become a reality—factory workers, warehouse workers, aides, attendants, and various admin assistant jobs. Anyhow, there are early indications that even some white-collar jobs may be affected.

To my mind, education should therefore focus on helping children develop their analytical skills, discover and enhance their creative talents, and teach them to not only follow procedures but, more importantly, gain skills to adapt to change and thrive in unpredictable environments. In addition, it may sound strange today, but education should also focus on training our children on basic human characteristics and personal interaction.

Why analytical skills? Well, while AI would be a great tool for data analytics, we would still need people to validate the analyses performed by AI machines. While AI would be able to analyze laboratory results, for example, and come up with diagnoses, we would still need humans to validate the diagnoses.

Why creative talents? Because creativity is one human characteristic that AI might be able to mimic but not inherently acquire. It is my opinion that creative works produced by humans would continue to be more valuable than those produced or generated by AI machines. And by creative talents, I just don't mean music, painting, or sculpture. I am also referring to culinary arts, carpentry, furniture design, interior decoration, architecture, landscaping, and many more creative disciplines. It is also possible that one day we would see technical jobs that require creative skills.

Why train our children on how to adapt to change and unpredictable environments? The application of AI, as I have pointed out in other sections in this book, is mostly targeted on functions that are logic-based and rules-driven in nature, meaning predictable. Therefore, the jobs that would almost certainly require humans would be those that would manage situations or perform tasks that could not be easily predicted by AI machines.

Education should also focus on the study of human characteristics of care, empathy, love, hope, etc. Humanity is on a disturbing trajectory. Technology has enabled global connectivity, yet a lot of people today feel so isolated. Millions of people claim to have many friends online, but the friendships are not deep when closely examined. Family members may live under one roof, but the individuals are glued to their mobile phones, computers, or video games—resulting in very little personal interaction.

Much more alarming is the amount and frequency of stimuli from electronic devices that our generation absorbs every single day. We humans are headed toward a future where we would be comfortable conversing and working with machines, but sadly, would lack empathy for each other. Because of this, I believe that there would be an increasing need for counselors, caregivers, doctors, nurses, psychologists, psychiatrists, and other mental health specialists as well as professionals providing human care and compassion.

Again, education needs to focus on the study of the basic human characteristics—for the sake of preserving our humanness.

Education needs to bring us back to the foundations of our humanity.

CHAPTER 12

TALK TO ME

One of our distinct characteristics as human beings is our ability to articulate our thoughts and feelings through words. In fact, a big part of our waking hours is spent on communication, wherever we may be. We communicate via email, text, or social media posts. We also communicate by way of engaging in a conversation—in person or over the phone.

Whether we do it via text or voice, we communicate to an individual or to a group of people.

TEXT-BASED COMMUNICATION

To effectively communicate through text, we need to use the right words, type the words correctly, and follow the grammatical rules.

Today's email technology is intelligent enough to make suggestions based on prior messages that you typed. For example, when you type, "This is to ack," the email system would autosuggest or autocomplete the sentence as "This is to acknowledge receipt of your email."

When typing your complimentary close in your email, you start with "Kin" and the system autosuggests, "Kind regards."

Clearly, there is some AI machine learning that is happening in the background. However, we are just seeing the beginnings of AI in text-based communication.

What would text-based communication look like with mature AI technology in place?

First, in choosing the right words, AI would be able to make suggestions when you ask the machine to give you some synonymous words that you could choose from.

Once you select the word you want to use, AI would perform a grammar check. If needed, AI would make suggestions on how to make the sentence grammatically correct.

Grammar checking, by the way, is no longer news. There is an AI-powered application in the internet called *Grammarly*[66] (www.grammarly.com) that could help you check for spelling and grammar.

Going forward, AI would be able to do more than that. If you are sending the text to a group of people, say your office colleagues, and need to send a formal letter, AI would be intelligent enough to format the letter following business letter-writing standards and make suggestions on how to make the words and sentences formal. Just like having a secretary to assist you. More advanced AI would be able to create draft letters with very little input from you.

Furthermore, AI could assist in translating your text to other languages, based on your instructions. For example, you are a global marketing executive and you need to send a letter to your customers who speak French, Mandarin, Spanish, German, Italian, etc. It is possible that AI, in the future, would be able to assist you to accomplish such task.

Imagine the efficiencies that such technology could offer, as well as the benefits it could deliver to people with disabilities.

VERBAL COMMUNICATION

The other method of communication is verbal communication.

There are a few challenges in verbal communication where AI could deliver value.

The first obvious challenge is people who are mute or have some form of speech disability. It is possible that science and technology would one day bridge that gap and have AI machines speak what a mute person has in mind. AI technology could also read out a text typed by a mute person.

Another beneficial use of AI in communication is in the field of speech-language therapy. AI equipped with natural language processing capability plus virtual reality would be a perfect tool for the development of speech-language therapy modules.

A "speech therapy aide" device could be useful for such treatment. But of course, speech-language pathology is much more than just helping a patient with speech sound articulation. Technologists and speech pathologists could work together to come up with an AI-enabled device that could be useful in speech-language therapy. This type of device would not only be beneficial to people with speech challenges but also assist elderly people who have dementia.

The second obvious challenge is language. People who travel abroad know this very well. When you travel to a country in Asia, you can have extreme difficulty navigating your way around. The same is true when you travel to a country in Europe where people might not be fluent in the language that you speak.

It is not remote that we would one day have AI devices that would break that barrier. It is possible that AI translators would allow you to speak into it (using an earphone with mic), say in English, and the AI translator would translate in real-time the sentences to another language, say French. The AI translator would then listen to the response from the French-speaking person and translate it to English through your earphone. Just like having a human translator with you.

That AI translator could help spur growth in travel and tourism. It could build bridges of friendship between people of different nationalities and languages who today would have difficulty communicating. It could help train people on how to speak other languages. It could even help songwriters write lyrics for their choruses in foreign languages.

The third challenge in verbal communication is understanding nonlinguistic signals—body language, tone of voice, activity level—that could indicate if the customer is listening, getting bored, or getting frustrated. Some people call this person-perception skill.

It is interesting to note that a company called *Cogito*,[67] founded in 2007 by Alex Pentland and Joshua Feast,[68] is applying AI machine learning technology to coach call center representatives while on the phone with customers—in real time!

Cogito measures timing of responses and audio activity level in a call center phone conversation to advise the call center representative whether the caller is actively listening, bored, or angry.

The fourth challenge in verbal communication, for a lot of people, is speaking before a group. Some people are suggesting that AI would one day be able to help humans prepare and practice their speeches. This application of AI would be useful to executives, politicians, and other public figures.

We have so far talked about the application of AI in text and voice communication. On the other hand, most of our communication today is done through our mobile phones and laptops.

AI ENHANCED DEVICES AND APPS

One interesting development to watch is the implementation of AI in mobile phones. We see traces of that today in that we can talk to our mobile phones and ask for, say, the nearest Spanish restaurant. As the application of AI on mobile phones matures, we would see more AI-enabled apps. That means apps that could talk to humans.

Take for example calendar alerts. Today, we get a ringtone and our mobile phones vibrate (if we have set it to vibrate) to remind us of an upcoming event in our calendar. An AI-enabled calendar could mean that the phone would speak to you and say, "Hey, John, don't forget that you have to call your mom. Do you want me to call her now?" or "It's time for lunch, Elizabeth. What food do you want to eat for lunch today?"

If you are driving, your AI-enabled mobile phone could answer your incoming calls by saying to the caller, "Christine is busy driving her car right now. Could I take your message?"

In case you are interested to hear the message from the caller, you could ask your mobile phone to play the voice message. You could then dictate your response to your mobile phone and have the AI app send the voice message as an audio file or text.

An AI-enabled app could also be developed and made available to help elderly people with dementia cope with the illness by reminding them of their medication, their upcoming appointment with the doctor, or even informing them what day it is.

Of course, it would be logical to have these AI voice features parameterized so that you could turn them off or on, depending on your circumstance. You would want to turn off the voice feature when you are about to enter a theater, a movie house, or a church.

Artificial intelligence apps could also enhance the security of your mobile phone so that it could confirm your identity as the owner through a combination of voice recognition, facial recognition, and finger-scanning technologies. This security feature is critical so that when you do any online banking transaction, travel booking, online shopping, or activity that would involve payment, your identity could be thoroughly authenticated and protected.

That said, with the application of AI technologies on mobile phones, all the online activities I mentioned above—online banking, travel booking, online shopping, etc.—could be done via text or voice. Adding VR technology in the mix would definitely make those online activities much more interesting.

As I have stated in previous chapters, imagine the benefits that these features could deliver to people with disabilities.

In the final analysis, the AI winter is over and the spring of man and intelligent machines has arrived.

CHAPTER 13

YOU GOT A FRIEND IN ME

So far, we have talked about AI in terms of "aide" devices, which could take many different shapes and forms. Such AI device could be a tubular appliance, a tablet, a piece of software running on a computer, a mobile phone app, an intelligent steering and auto navigation system, or an advanced logistics system controlling a fleet of equipment. It could also be some algorithm running on your virtual reality headset or your home entertainment system.

However, artificial intelligence has always been associated with machines that walk like humans, move like humans, and talk like humans—machines we call *robots*.

While robotics technology has been used in the space industry for many years now, this technology has finally landed on earth. There are now robots or androids that have faces like human beings, eyes that "see" and move like human eyes, ears that "listen," and brains that "understand" and carry a conversation. Some robots even have the agility of humans. In other words, they can walk, jump, run, carry weight, stand, sit, etc.

Boston Dynamics, a spin off from Massachusetts Institute of Technology, has developed a robot called *Atlas* that is dubbed as "The World's Most Dynamic Humanoid." Below is an excerpt from their website:

Atlas is the latest in a line of advanced humanoid robots we are developing. Atlas' control system coordinates motions of the arms, torso and legs to achieve whole-body mobile manipulation, greatly expanding its reach and workspace. Atlas' ability to balance while performing tasks allows it to work in a large volume while occupying only a small footprint.

The Atlas hardware takes advantage of 3D printing to save weight and space, resulting in a remarkable compact robot with high strength-to-weight ratio and a dramatically large workspace. Stereo vision, range sensing and other sensors give Atlas the ability to manipulate objects in its environment and to travel on rough terrain. Atlas keeps its balance when jostled or pushed and can get up if it tips over.[69]

Perhaps the most famous robot today is called *Sophia*, a social robot developed by Hanson Robotics and designed to be a companion for humans including the elderly. Sophia is an android that looks, acts, and talks like a human being. The software behind this robot was designed to learn over time and make it increasingly more intelligent. Sophia has delivered a speech to the United Nations and has been given citizenship by Saudi Arabia.[70]

There are now robots that are designed to assist you in your kitchen. They are programmed to cook a wide variety of dishes for you. Best of all, these robots can also wash dishes.

There are also robots that are designed to be "pets" like dogs, cats, birds, and snakes. I am not quite sure why one would prefer a robot pet over a real pet. Nonetheless, there is very significant reason why this specialized area of robotics is not to be ignored. Think about the potential of being able to miniaturize these kinds of robots. These intelligent machines could be used to probe places that humans have never been able to explore before, such as caves, geophysical cavities, trenches, pipes, termite or ant colonies, and deep sea. If we could miniaturize them enough, we could even use them to probe the human body.

In other words, these robots could go where we people could not. They could be used for rescue operations, scientific explorations, and even antiterrorist activities—without risking human lives.

The current generation of robots is quite expensive. But the good news is that just like any commodity, they would become cheaper as their market grows.

The growth of the robotics industry would create a new set of jobs: robotics strategists, business analysts, robot designers, robot testers, sales and marketing executives, robot installers or implementors, and of course, maintenance and support people specializing in robotics.

Robots carrying boxes.

When that happens, robots would be a part of our daily routine. We would have robots working in the office, robots carrying boxes in a warehouse, robots helping with house chores, robots at grocery cash registers and bagging areas, robots in schools, robots guiding elderly people or children across the street, and robots taking orders in a restaurant.

It could also mean robots conversing with elderly people in a home for the aged. It could also be robots watching over hospital patients. Or robots assisting disabled persons.

What this possibly means is that a considerable number of people who are currently doing these functions could lose their jobs.

Some would argue that robots taking over human jobs would not be a good thing, therefore, manufacturing robots is not ethical. It is a reasonable speculation that a considerable number of people think that robots are our enemies and that we should stop creating them. But the same line of argument could be used against automated teller machines, airline check-in kiosks, and many other technologies that deliver cost savings and efficiencies to businesses.

The fact is that regardless of ethics, robotics engineering and development will advance. This highly specialized industry vertical will continue to research and improve its products and services for as long as there is opportunity for business. As a result, decades from now, robots will become sophisticated.

When we reach that point, some could say that we live in the company of strangers. But then are we humans not slowly becoming strangers to one another? I do not wish to get into a philosophical discussion about this. But who knows if robotic technology could one day fill that widening gap amongst us humans, who are slowly drifting apart from each other?

The ultimate question is this: "Are robots our friends or foes?"

The answer: "Only time will tell."

CHAPTER 14

REMEMBER MASLOW

Food is essential to humans. So is water. Without food and water, people will die of starvation and dehydration in a few days.

While there are countries where food and water abound, there are also places that are struggling; their people are dying of starvation and dehydration by the thousands per year.

Food production needs water. Not only water but also a good environment for the plants to grow and for the animals to live and propagate.

It is also important to note that plants are not all the same. There are plants that grow in mountainous areas, in low land, and in special conditions like hot and sunny weather, while other plants like less sunlight.

Correspondingly, animals are not all the same. Some animals need grass, some feed on corn or rice, and some thrive in warm weather while some prefer a cold habitat.

Clearly, we need to factor in seasons when we plan food production cycles. There is a time to plant and a season to harvest.

For food to reach people, transportation is required. And transportation needs infrastructure (i.e., roads and bridges), gas, and electricity (in the case of electric powered vehicles).

As for distribution, any person with experience in logistics would tell you that there is something called the "optimal route" to distribute packages or goods.

As you can see, there are many factors to cogitate when planning food production and distribution. There is a need for an integrated plan that takes into account all the aforementioned factors.

Planning, as described above, would be critical even for a place or country where there is an abundance of food and water. But for countries that do not have that luxury, proper planning could spell the survival of its people.

How could artificial intelligence help in this scenario?

The strength of AI is in ingesting large amounts of historical information and learning from that to solve future problems. In this case, AI could be used to establish patterns from historical data involving weather, cycles of water supply, planting in multiple regions, harvesting, cycles of animal reproduction (i.e., cows and chicken), and geographical distribution of population—to help humans plan food production and distribution.

Though I mentioned above that such comprehensive planning is crucial even to countries with rich resources, this ability to plan in detail—taking into consideration numerous variables—would likewise be useful at a microscopic level. This refers to every business that is part of the food supply chain and logistics.

An interesting development to watch is the application of robotics in the food production and distribution workflow.

On May 22, 2018, an article entitled "How Robots and AI Are Transforming the Food Industry" was published on the Dell Technologies website. Below is an excerpt.

> "There has been a lot of growth in the use of robotics and automation in the food industry," Bob Doyle, vice president of the Robotic Industries Association, said. Although much of this technology was developed for the automotive industry, more nontraditional

industries, including food and beverage, are beginning to adapt, employ, and benefit from these types of robotics, data, and vision systems. Part of the reason, Doyle explained, is that the technology—which was originally developed to pick up sturdy metal parts—has become sophisticated enough to pick up fragile items such as tomatoes, eggs, and loaves of bread without destroying them.

In addition to automating repetitive tasks, Martial Hebert, a professor with the Robotics Institute at Carnegie Mellon University explained that robotics and artificial intelligence (AI) are enabling food manufacturers to track products and consumer demand—then match production to this demand, based on data analysis. When companies are better able to analyze important steps like production, transportation, and refrigeration, he said, they're also better able to monitor food safety, including whether food was inadvertently contaminated and where that food was shipped and sold.

In the last few years, several food producers made hefty investments in robotics. CMC Food, an egg producer in Fanwood, New Jersey, for example, built a new production facility around automation. In the factory, CMC Food replaced human egg handlers with two robots able to manage more than 100,000 eggs per hour.

"In the past, the workers would have to feed 10 dozen eggs at a time into the machine and also stack the finished boxes at the end," Michael Culley, CMC Food president, said in a CNBC article. "Now the robots handle the heavy parts."[71]

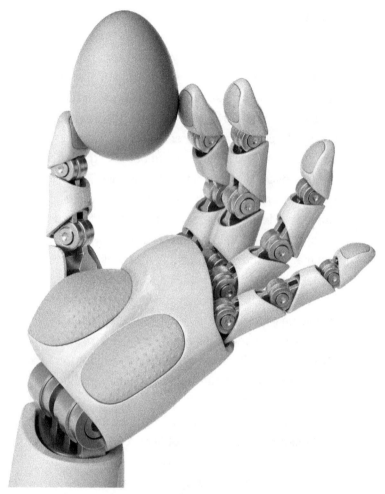

Robot hand holding an egg.

There are also remarkable developments in the application of robotics in the warehousing industry, which is an integral part of food distribution.

Boston Dynamics specializes in creating robots for warehousing operations. These robots play a major role in the handling, storage, and distribution of perishable goods—such as meat, fruits, and vegetables—as well as non-perishable items.

A.I. HACKED

On April 2, 2019, Boston Dynamics released the following news article on their website:

> Boston Dynamics today announced the acquisition of Kinema Systems, a Menlo Park-based company that enables industrial robotic arms with deep learning technology to locate and move boxes on complex pallets. Using a combination of vision sensors and deep learning software, Kinema Systems' Pick technology works with leading commercial robotic arms to move boxes off pallets to conveyors or build stacks of boxes on pallets. Pick enables logistics, retail, and manufacturing companies to achieve high rates of box moving with minimal set up or training for both multi-SKU and single-SKU pallets. [72]

So, for AI machines to be able to plan, forecast, and in the case of robots, handle objects as delicate as eggs, it means one important thing to the food industry: minimal downtime. Machines do not resign, machines do not sleep or take vacations, and they have no behavioral issues.

I would like to pause here and let you ponder this: if mankind will compete against machines that pose no risk of leaving the company, machines that do not sleep, machines that do not take vacations, and robots that have no behavioral issues, then how should we humans strategize for this upcoming competition?

CHAPTER

THE SCIENTIFIC WAY

I f there is one area that would hugely benefit from AI, it would be science.
I am talking of research for new medicines. I am referring to improved efficiencies and accuracy in the provision of patient care. I am speaking of enhanced precision in surgery and diagnostic medical imaging. I am also talking of acceleration in scientific discoveries.
Let's take medicine, for example.

ACCELERATING DRUG DISCOVERY

Top drug companies like AstraZeneca, BASF, Bayer, Bristol-Myers Squibb, Celgene, GSK, Janssen, Lilly, Merck, Novartis, and Sanofi are working with AI companies to accelerate research for new medicines.
Case in point, Sanofi is working with a company that has an AI platform that could learn from massive repositories of data on drug discoveries. They are using AI to discover bispecific small molecules to cure metabolic diseases like diabetes and its comorbidities.
On May 9, 2017, the *Genetic Engineering & Biotechnology News* website released an article entitled "Sanofi, Exscientia Ink Up to €250M Deal for Bispecific Drugs Against Metabolic Diseases". Below is an excerpt from that article:

> Sanofi and Exscientia signed a potentially €250 million (approximately $273 million) collaboration and license option deal to discover bispecific small-molecule drugs against metabolic diseases. Scotland-based Exscientia will use its artificial intelligence (AI)-driven platform and automated design capabilities to identify combinations of synergistic drug targets, and then apply its lead-finding platform to identify bispecific small molecules against those targets.
>
> ...
>
> Exscientia's drug discovery engine is founded on an AI platform that the firm claims can learn best practice from huge repositories of existing drug discovery data. The platform allows the design and evaluation of novel compounds for predicted criteria, including potency, selectivity, and ADME, against specified targets.[73]

Meanwhile, we now have a complete map of the human DNA. From this map, we know what DNA type is more predisposed to certain diseases like cancer. Not only have we achieved that, but we also have identified the DNA component that is responsible for aging!

In 2009, the Nobel Prize for Physiology or Medicine was awarded to three individuals—Elizabeth H. Blackburn, Carol W. Greider, and Jack W. Szostak—who found the DNA component responsible for aging. The following appeared in the Nobel Prize's official website:

> This year's Nobel Prize in Physiology or Medicine is awarded to three scientists who have solved a major problem in biology: how the chromosomes can be copied in a complete way during cell divisions and how

they are protected against degradation. The Nobel Laureates have shown that the solution is to be found in the ends of the chromosomes—the telomeres—and in an enzyme that forms them—telomerase.

The long, thread-like DNA molecules that carry our genes are packed into chromosomes, the telomeres being the caps on their ends. Elizabeth Blackburn and Jack Szostak discovered that a unique DNA sequence in the telomeres protects the chromosomes from degradation. Carol Greider and Elizabeth Blackburn identified telomerase, the enzyme that makes telomere DNA. These discoveries explained how the ends of the chromosomes are protected by the telomeres and that they are built by telomerase.

If the telomeres are shortened, cells age. Conversely, if telomerase activity is high, telomere length is maintained, and cellular senescence is delayed. This is the case in cancer cells, which can be considered to have eternal life. Certain inherited diseases, in contrast, are characterized by a defective telomerase, resulting in damaged cells.[74]

For a while now, there have been discussions on developing drugs that are DNA specific. AI could help accelerate that kind of research. AI could help humans analyze drugs and DNA compatibilities. Highly sophisticated AI algorithms could sort through thousands of research papers and medical journals to find patterns that could lead to valuable information. Without the aid of AI, it would perhaps take humans years if not decades to complete those researches.

AI could also help find patterns by scanning through thousands of medical records containing patient DNA, medicines prescribed, and outcomes. AI could establish correlations leading to new protocols in managing a wide array of diseases and DNA types.

Due to the acceleration in the development of DNA-specific drugs, it is possible that the life expectancy of human beings would someday exceed one hundred years. When we reach such point, a new set of challenges would arise covering retirement, social security, health insurance or Medicare, food and water supply, shelter, etc.

SMART MEDICAL DEVICES

As far as the future of patient care is concerned, it is also possible that doctors would one day be aided by AI machines that would suggest possible diagnoses or remind doctors of dosage thresholds, side effects, allergic reactions, and drug contraindications. That said, researchers are now tapping into AI to develop "smart medical devices".

On February 5, 2019, an article appeared on the *Digital Trends* website (www.digitaltrends.com) stating that Johns Hopkins University researchers have designed a device that uses AI technology to help automatically screen for pneumonia by listening for particular types of breathing on the part of patients. As a result, the stethoscope itself can provide diagnoses.[75]

We could expect many more AI-enabled medical devices in the coming years. For example, AI could also help nurses *monitor the monitors* and call on the appropriate specialist or doctor on a case-by-case basis.

Is it then possible for AI machines to completely take over the work of nurses? I doubt that it would, though the impact of AI might be such that hospitals would require fewer nurses.

How about the use of AI in surgery?

Surgery is one specialized area of medicine that could greatly benefit from AI. There are now companies, such as Galen Robotics, that specialize in developing robots to assist surgeons in realizing precise minimally invasive surgery. Below is an excerpt from their website:

> Galen is developing a single-platform solution to aid surgeons across several disciplines with minimal

disturbance to existing workflows. Our cooperative control paradigm aims to eliminate hand tremor and enable surgeons to realize precise minimally invasive interventions in otolaryngology, spine surgery, and tissue reconstruction that were previously considered beyond human capacity. [76]

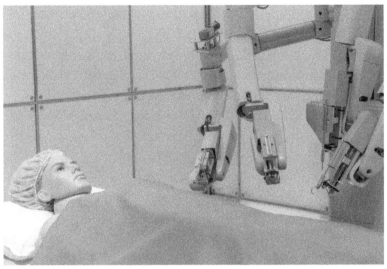

Experimental medical robotic surgery.

Due to the fact that robots are controlled by electronics, this could also mean that surgeons would be able to perform their high-precision surgery from remote locations. This capability to perform surgery remotely would be very useful in emergency situations where there is no time to fly in a surgeon. This technology would be vital to the military. For example, there is need for a surgeon located in Maryland to perform remote surgery on a soldier deployed in the Middle East.

Remote surgery is no longer fiction. The first successful and highly celebrated remote surgery happened on September 7, 2001. It was dubbed the *Lindbergh Operation*—named after Charles Lindbergh, who was the first person to fly a custom-built, single-seat monoplane across the Atlantic. The Lindbergh Operation was performed by French surgeon Dr. Jacques Marescaux, in New York City, on a sixty-eight-year-old patient in Strasbourg, France.[77]

How about AI in diagnostic medical imaging?

Artificial intelligence is sweeping through the field of diagnostic medical imaging. An article written by Shiddharth Shah and Robin Joffe entitled "How AI Is Evolving in Diagnostic Imaging" (www.diagnostic.com), released on December 10, 2018, stated that out of 114 startups, "a significant majority target the image analysis aspect of medical imaging.".[78] A radiologist's job is to identify and analyze specific characteristics in an electronic diagnostic image. This function is critical in the overall provisioning of medical care.

Such application of AI would not only be useful in first world countries but would also greatly benefit areas around the world where there is a shortage of highly skilled radiologists.

AI IN SCIENCE AND TECHNOLOGY

How about scientific discoveries?

The application of AI in scientific research would yield a lot of benefits.

AI could be applied in global climate science, environmental science, marine science, materials science, chemistry, geodetics, physics, electronic and electrical engineering, civil engineering, aviation, astronomy, and even space exploration. Robotics is already applied in some of these areas.

The use of AI in science would lead to the acceleration of new inventions, new scientific discoveries, new standards in various fields of engineering, and a host of other benefits.

AI could assist humans in the management of energy, gas, and water supply for cities and states.

In the case of energy management, *MIT Technology Review* released an article about a company called *Numenta* (www.numenta.

com)[79] in February 2013. Below is a summary sentence from the article entitled "Numenta's Brain-Inspired Software Adds Smarts to the Grid."

> The company's Grok software processes "fast data" for EnerNoc and makes predictions about customers' energy usage.[80]

AI could further assist city or state governments in planning and providing essential services to their constituents. This could be achieved by offering analytical insights, taking into account factors like age, average life expectancy, income level, geographic spread of population, migration trends, location and capacity of hospitals, transportation, among many others.

AI could also help manage vehicular traffic for an entire city. AI could ingest thousands or millions of traffic flow data pieces from numerous monitoring points across a city and program the traffic lights accordingly. This could result in a more efficient flow of vehicles. And if you add the idea of unmanned vehicles operated by intelligent electronics that could communicate with the traffic control system, you could potentially solve the world's traffic problems.

In summary, the application of AI could benefit not only the developed countries but also the developing nations in the area of health care, science, engineering, and even city traffic management.

AI could help solve humanity's biggest challenges in a much more efficient and expeditious manner. Unlike the human brain, AI machines can absorb millions upon millions of pieces of data and derive patterns to solve large and complex problems.

So, is AI a friend or a foe?

CHAPTER 16

JUSTICE DELAYED IS JUSTICE DENIED

The law practice is one profession that comes close to the medical profession in terms of volume of information.

Sifting through a massive repository of decisions rendered by the highest court of the land, the United States Supreme Court, as well as lower courts, is undoubtedly a daunting task even for a team of brilliant lawyers.

For each litigation case, the artifacts that need to be reviewed could reach thousands, if not millions, of files. These files could be in the form of audio recordings, documents, faxes, text messages, and email messages. The amount of time and resources involved in gathering, reviewing, labeling, grouping, and analyzing the artifacts could take many months.

Today, there are companies like Brainspace (www.brainspace.com), Everlaw (www.everlaw.com), and Relativity (www.relativity.com) that offer electronic discovery solutions utilizing machine learning, analytics, and visualization technologies. Their software solutions provide government agencies, corporations, law firms, and even consulting companies the ability to manage files relating to litigation, investigation, and Freedom of Information Act (FOIA)[81] requests. These AI-enabled platforms are capable of ingesting large volumes of data (in various file formats) and organizing them for subsequent search, review, and analyses by lawyers, investigators, analysts, and other authorized users.

These systems are intelligent enough to establish word correlations and contextual associations like "Independence Day" to "Fourth of July" and "Fourth of July" to "Freedom."

Fast-forward into the future. As AI and robotics technologies mature, it is possible that robots would find their way into the investigation and litigation workstreams. This could mean AI machines or even robots doing research for human lawyers by way of scrutinizing volumes upon volumes of court decisions.

Furthermore, these AI machines or robots could collect and compile other related artifacts, review them, perform some preliminary analysis, and present their findings to a human lawyer, investigator, or law enforcer.

With such advanced AI technologies in place, we could expect that investigations and litigations in the future would be expeditious.

However, the same AI technology that could make lawyering more efficient could also make it complicated. Think about it. The matter of intelligent machines that could have their "own minds" would give rise to a multitude of issues.

Consider the following:

- Is a robot a distinct entity from its creators?
- If an unmanned car accidentally hits a person, who is liable: the owner of the car or the manufacturer of the "intelligent" car?
- If a robot becomes violent, who bears the responsibility? Is it the manufacturer or the owner/buyer of that machine?

In the same vein, if that AI machine "learns" to buy a lotto ticket and wins the lotto, who gets the prize money?

It may sound amusing, but the implications of AI in terms of laws, regulations, insurance claims, and liabilities are far-reaching.

All things considered, it is my belief that the law profession would survive the future, though the jobs of human clerks and researchers employed by law firms could be affected by the application of AI technologies.

Now, think about this: given two law firms having access to the same AI technologies, where would the competitive advantage lie?

The competitive advantage would remain in the hands of skilled lawyers who could expertly debate and defend their clients.

In the final analysis, when it comes to the field of justice, humans would continue to lord over the land.

CHAPTER 17

A MIGHTY FORTRESS

Security is of prime importance to any individual, family, neighborhood, society, city, state, nation, or region.

The laws of the land are legislated to protect the country's borders and territory, all the way down to its individual citizens. Every nation builds its military and intelligence capabilities to protect itself from external threats. And every country, state, city, or municipality, has its respective law enforcement groups to ensure that the laws and regulations are adhered to by its constituents.

The strategic elements that underpin security—covering geographical, social, economic, and political spheres—include treaties, negotiations, weapons, military and security personnel, bases, equipment and supplies, logistics and defense capabilities (ground, sea, and air), and a host of advanced technologies.

The military is known to have invented most of the technologies that we know today, including wireless communication, Global Positioning System, night vision, and cryptography.

It sounds logical then that artificial intelligence technologies, including robotics, should be implemented in the military and defense industries. The benefits are immense: efficiency, accuracy, cost savings, etc.

Before we all get too excited about it, we need to be reminded that computer security is perhaps one of the most challenging areas in the world of security—whether it is at the national, state, city, or individual level.

Case in point, the number of computer security breaches from 2005 to the present is not decreasing, despite the increasing number of new security solutions that are being spawned every year.

According to the website (www.statista.com) of Statista, the leading provider of market and consumer data, the number of data breaches reported in 2017 peaked at 1,579 incidents, compared with 157 incidents in 2005. The total number of records exposed in 2017 was reported at about 179 million. Those numbers are in the United States alone.[82]

It is said that the reason for hacking computer systems has transformed from mere "bragging rights" to financial gain. Criminals have now realized that they could use technology to steal personal records and sell them. They have figured ways to steal money without physically robbing a bank. They have found ways to compromise the security of corporate systems to demand for ransom money. They have learned how to spoof the identity of individuals to be able to charge expenses to unsuspecting persons.

We have even seen news about some of our systems that were allegedly hacked by another country to influence our elections.

I would not even go down the path of speculating a scenario where a defense system, which runs on artificial intelligence, would one day have a mind of its own and target us.

The mere specter of a rogue AI code or virus that could hand the controls of our defense systems to a foreign country, group, or individual, should be enough reason for us to think many times over before going forward with applying AI in that critical area of our national security. Nonetheless, it makes sense to apply AI in military and defense because if we do not, then our technological capabilities would be way behind that of other countries.

The ultimate question then is this: Should we equip our military systems with AI or not?

The floor is now open for discussion.

CHAPTER 10

WHAT MAKES THE WORLD GO 'ROUND

Let's talk big picture.
Most countries measure the health of their economies by numerous factors, including the number of new jobs created each month. In the US, we go by an average of 200,000 new jobs every month to be able to say that the economy is healthy.[83] When that figure declines, experts start to worry about a possible recession. Recession means that the economy is slowing down.

When the monthly stats on new jobs are in a downward trend, that means that businesses are cutting back on hiring, which translates to businesses not growing. And when businesses are not growing, the economy is shrinking.

Why is that? Well, because if people do not have jobs, then people would not earn and would not have money to spend. And when people do not spend money, then the sales of businesses would be affected. And if the revenues of businesses shrink, then taxes would go down too. And if the taxes collected are not growing, then the wealth of the nation is shrinking. I know this is too simplistic, but it roughly describes how the economy works.

We need the wealth of the nation to grow because we need the government—the managers of the country's wealth—to continue to provide vital services to the public. We need the government to build roads, bridges, schools, hospitals, and day care centers for the elderly, maintain parks, provide Medicare, etc.

Now, if we aggregate all the economies around the world—economies of both rich and developing countries—then we would have a picture of what we call the *global economy*.

We say that the global economy is growing if the sum of all GDP (gross domestic product) of all countries around the world is going up year on year. The GDP of some countries may shrink, but as long as the combined GDP of all countries is increasing, then the global economy will still be said to be growing.

How does this relate to artificial intelligence?

Well, AI has been become such a hot topic in economic circles that big global consulting companies have analyzed the impact of AI on the global economy.

The result?

PricewaterhouseCoopers (PwC), which is one of the largest consulting companies in the world, estimates that "AI will add $15.7 Trillion dollars to the global economy."[84] Their forecast is that the global GDP would be 14 percent higher in 2030 because of artificial intelligence, according to *Bloomberg* on June 28, 2017. That's in 2030—just a few years from now.

What this means is that experts are predicting a rapid growth in the application of artificial intelligence in the coming years. We have now reached a point of no return.

In view of the inevitable, how should you prepare for the future? How should you guide your children and your children's children? The reality that they will experience a few decades from now would be nowhere near the world that we see today. Theirs will be a realm of humans and intelligent machines.

While we hear people say, "Two heads are better than one," technologists are now saying, "Man and machine together is better than the human."[85] That's where our world is headed.

Regardless of how the rest of humanity looks at AI machines, businesses will use AI technology—anything or any technology for that matter—to enable them to realize profit and achieve growth.

Finally, let's not forget that the growth of businesses is a major contributor to the growth of our national economy. And the sum of all national economies is what makes up the global economy.

And that is what makes the world go 'round.

CHAPTER

THE NEXT FRONTIER

As I pondered the last chapter of this book, over a billion people around the world were celebrating Chinese New Year.

I thought that it would be fitting to go to a famous dim sum restaurant in Manhattan's Chinatown and have a sumptuous lunch. It was there that I serendipitously shared a table with a visiting Hungarian marketing strategist in the fashion industry. Yes, by the way, it was also New York Fashion Week.

Our conversation somehow drifted to artificial intelligence. Just like most of the people I talked to about AI, she expressed concerns about people losing their jobs. While I told her that I share the same concerns, I explained that the AI industry would create new jobs that would require new skills. She then asked what advice I would give to people in order for them to prepare for the future.

So here are my thoughts.

As I discussed in prior chapters, artificial intelligence technologies would most likely be applied in areas where the job functions are repetitive, logic-based, rules driven, and predictable. These would be clerical work, administrative functions, labor-intensive jobs—often referred to as blue-collar jobs. Even white-collar jobs, in the case of radiologists, could be affected too. We discussed in chapter 14 how robots are now replacing human egg handlers in an egg production plant.

So, what skills would then be valuable in the future?

The jobs that people would be competing for in the future would require interpersonal skills, analytical skills, technical proficiency, creative abilities, and the competence to handle or manage unpredictable situations or conditions.

First, here are some of the jobs that would likely survive the future, noting that they would require interpersonal skills: sales and marketing, patient care, hospitality, counseling, therapy, jobs that specialize in negotiations, etc.

Second, here is a list of some jobs—which require analytical skills—that would continue to be in demand in the future: data scientists, technology architects, science or medical researchers, doctors, lawyers, managers, executives, etc.

Third, the AI industry would also create demand for a new breed of technical people who specialize in the field of artificial intelligence: AI business strategists, business analysts, architects, testers, installers, IT security, and maintenance and support experts. There would also be a need for technical people in the following areas of data management: data preparation, data transformation, data quality, data governance, etc.

Fourth, it is probable that jobs requiring creative skills would continue to be valuable: musicians, painters, sculptors, visual artists, architects, designers, writers, novelists, culinary experts, animators, filmmakers, etc.

Finally, the jobs that would likely survive the future would require the skill to adapt to unpredictable circumstances or conditions. These are the strategists, managers and executives, project managers, physicians, airplane pilots, disaster and emergency response personnel, security professionals, military and law enforcement specialists, etc.

Though the skills I mentioned above would be in demand, there is no guarantee that there would be enough jobs for everyone who possesses those abilities. Clearly, competition for work would be steep.

All the jobless people would need to be supported by the

government for humanitarian reasons as well as for the country to avert an apocalyptic plunge into an abyss of irreversible economic depression.

But whatever happens to the job market, AI technologies would continue to march forward.

So what would the future of this technology look like?

As I contemplated this question, I chanced upon a video clip in social media of a man with prosthetic arms. At a closer look, I noticed that the man could control not just his prosthetic arms but also his fingers. It took me a moment or two to realize that his prosthetic was actually being controlled by his mind. It reminded me of a TV series in the 1970s called *The Six Million Dollar Man,* which featured a cyborg who had extraordinary strength due to bionic implants in his body.

While the video clip was not exactly an equivalent to a bionic man, this advancement in biomedical engineering is no less remarkable.

Come to think of it, in just about four decades, we have gone from pure fiction to reality.

This fascinating mind-controlled prosthetic was developed in the Johns Hopkins Applied Physics Lab (APL) by way of "brain mapping" or observing how the human brain works. This means determining where neurons are fired in the human brain when a person raises his left arm, or right arm, or both, or when a person moves each finger, or moves all fingers to grab a cup of coffee. They were then able to program the prosthetic to move depending on which part of the brain was firing electrical signals. Here is an excerpt from an article in the *Johns Hopkins Health Review* website entitled "Mind-Powered Prosthetic Limb":

> The brain prosthesis interface that APL developed works via electrodes on the skin detecting the tiny electric signals that occur whenever a muscle is activated. If you want to move your index finger, for instance, your thoughts trigger muscles in your

forearm, which in turn activate the correct tendon to move the finger. With the Modular Prosthetic Limb, electrodes pass along these muscle-movement signals to a computer within the prosthesis, which interprets them to activate the correct motors to move the hand and fingers.[86]

It is one thing to talk about machines or some electronics as independent entities—separate from human beings—and another thing to talk about them as part of the human body.

As I further contemplated the video, there was one other possibility that astounded me.

In case we haven't done so already, when we reach a point where we have a complete map of the brain—that is, we know which part of the brain controls what part of our body—and we could create electronics or machines that could translate our brain signals and execute them into action, then we have paved the way for humans to completely control machines or robots.

The implications are far and wide.

We could deploy robots anywhere on this planet and have them controlled by human brains from a remote location. Such technology could be used for military purposes, for scientific probes, for space explorations, or just about any good reason we could think of.

Add virtual reality technology into the mix and the human who is controlling the robot from another location would be able to visualize what the machine is going through.

While we may find the above scenario fascinating, there is another sector of artificial intelligence architects who have been studying how the human brain works to design and implement deep neural networks—electronic networks that operate like the human brain.

A company called *Numenta* is now facing this challenge head-on. Here is a blurb on Numenta's website:

A.I. HACKED

> Numenta is tackling one of the most important scientific challenges of all time: reverse engineering the neocortex. Studying how the brain works helps us understand the principles of intelligence and build machines that work on the same principles. We believe that understanding how the neocortex works is the fastest path to machine intelligence, and creating intelligent machines is important for the continued success of humankind.[87]

If I am to further stretch my imagination, how awesome would it be if we are able to translate the human brain's neural signals to digital format—in the same way that we were able to digitize music and video. I am not even sure if this is possible.

But if we could digitize the human brain's electrical signals into bits and bytes—in a way that intelligent machines could seamlessly integrate with human brains—then we would have reached the next frontier in artificial intelligence: man-machine interface. A link through which machines serve as extensions to human brains. Ray Kurzweil, an American inventor and futurist, calls it *Singularity*.[88]

Imagine the powerful computing capability and massive memory capacity that such interface could add to the human brain.

If we ever achieve that feat, there is something even more astounding that we humans would have consummated: the cerebral union of the entire humanity and the global network of machines.

If we would ever see that day come, then at that epochal point in the history of humankind, machines would no longer operate on artificial intelligence. They would operate on true human intelligence!

INDEX

A

A.I. x, xi, xii, 5, 6, 8, 9, 10, 11, 12, 13, 14, 15, 17, 19, 21, 23, 25, 26, 30, 31, 32, 34, 35, 37, 39, 44, 45, 48, 49, 50, 51, 52, 53, 54, 56, 57, 58, 59, 61, 62, 63, 64, 65, 67, 72, 75, 77, 79, 80, 82, 83, 85, 86, 87, 90, 92, 95, 96, 97, 106, 107, 108, 109
A.I. winters 6
Alan Turing 5
Alexa x, 105
algorithms xi, 8, 9, 10, 13, 25, 31, 45, 49, 79, 106
Allen Newell 5
Amazon ix, x, 105, 106
analytics 15, 52, 59, 85, 108
Art 29
Artificial Intelligence x, xi, 5, 8, 9, 10, 11, 12, 15, 23, 31, 39, 49, 52, 58, 65, 67, 72, 82, 89, 90, 92, 95, 96, 98, 99, 105, 106, 107
artwork x, 32, 33, 34, 35, 109

B

Beethoven 33, 109
Bloomberg x, 105, 112
Body Chip 46
Brainspace 85

C

Chopin 33, 109
Christie's x, 32, 34, 105
CMC Food 73
copyright xi, 29, 30, 31

D

data 6, 7, 8, 9, 10, 12, 31, 37, 38, 39, 45, 46, 50, 51, 52, 53, 59, 72, 73, 83, 85, 90, 96, 106
Deep Learning 8
digital marketing 42
DNA xi, 5, 43, 78, 79, 80, 106

E

Everlaw 85
expert systems 7

F

Facebook x, 105
Freedom of Information Act 85, 111

G

Google x, 17, 106
Google Duplex 17

H

Hanson Robotics 68, 110
Herbert Simon 5
hospitality 51, 96

I

IBM Watson ix, 105
Impressionist movement 33
insurance xii, 51, 54, 80, 86
intellectual property xi

J

Jeopardy ix, 105
jobs 15, 49, 50, 52, 53, 58, 59, 69, 70, 87, 91, 95, 96
Johann Sebastian Bach 29
John McCarthy 5
Johns Hopkins University 80

L

LA Times 47, 110
logistics 39, 51, 67, 71, 89

M

machine learning xi, 8, 9, 61, 85, 106
machine perception xi
Machine perception 8, 106
Marvin Minsky 5
medical imaging 82
Michelangelo 33, 109

N

natural language processing xi, 9, 12, 26, 63, 106, 108
neural networks 98
New York Times 43

P

Pablo Picasso 33
Planning 9, 72
Portrait of Edmond Belamy 32
PricewaterhouseCoopers x, 92
Production Aide 39
prosthetics 97

R

Relativity 85
Robotic Process Automation 10, 108
robotics xi, 50, 67, 68, 69, 72, 73, 86, 89
robots 9, 67, 68, 69, 70, 73, 75, 81, 86, 95, 98, 106, 110
royalties xi, 31

S

security xi, xii, 44, 46, 47, 50, 51, 53, 65, 80, 89, 90, 96
social media 12, 13, 26, 38, 41, 42, 105
stethoscope 80, 111

T

Teaching Aide 57

telomere 79
Tesla x, 106
translator 63

V

Virtual Reality 37, 39, 57, 63, 67

NOTES

INTRODUCTION

1. IBM Watson is IBM's artificial intelligence platform.
2. *Jeopardy* is a quiz competition on television. The show was created by Merv Griffin and is hosted by Alex Trebek.
3. The *Telegraph* article can be found in the following link: https://www.telegraph.co.uk/technology/2016/01/28/first-driverless-buses-travel-public-roads-in-the-netherlands/.
4. Link to the *MIT Technology Review* article: https://www.technologyreview.com/s/601339/will-artificial-intelligence-win-the-caption-contest/.
5. Link to the *Forbes* article: https://www.forbes.com/sites/eustaciahuen/2016/10/31/the-worlds-first-home-robotic-chef-can-cook-over-100-meals/#638a5327228b.
6. Link to the *Quartz* news item: https://qz.com/875491/japanese-white-collar-workers-are-already-being-replaced-by-artificial-intelligence/.
7. The *Bloomberg* article can be found in the following link: https://www.bloomberg.com/news/articles/2017-06-28/ai-seen-adding-15-7-trillion-as-game-changer-for-global-economy.
8. The article on Christie's Auction House website can be found in the following link: https://www.christies.com/features/A-collaboration-between-two-artists-one-human-one-a-machine-9332-1.aspx.
9. Facebook is a social media platform on the internet for sharing words, photos, and videos with friends.
10. Amazon Alexa, also known as Alexa, is a virtual assistant developed by Amazon, which also owns the Amazon online shopping portal.

11   Google Home is a speaker and voice assistant. This product is manufactured by Google, the same company that developed the famous internet search engine.
12   Tesla is an automotive and energy company based in Palo Alto, California, that specializes in electric car manufacturing. Tesla was founded by Elon Musk. Here is a link to a *Forbes* article on Tesla's autodriving cars: https://www.forbes.com/sites/jeanbaptiste/2018/11/07/tesla-could-have-full-self-driving-cars-on-the-road-by-2019-elon-musk-says/#649099e962ac.
13   Amazon Prime is a subscription-based online shopping portal.
14   A chatbot is a robotic software designed to chat with a client for purposes of customer service or provisioning of information to customers.
15   Machine perception is subdomain of artificial intelligence that aims to provide computer systems the capability to interpret data in a way that humans use their senses.
16   Facial recognition and voice recognition are capabilities of biometric AI systems that are used to identify or authenticate a human being.
17   Machine learning is a subdomain of artificial intelligence that is involved in developing computer algorithms to train machines to learn like human beings.
18   Natural language processing is a subdomain of artificial intelligence that is concerned with enabling computers to process and analyze natural language (i.e., human language) data.
19   Robotics is a subdomain in artificial intelligence that focuses on the development of physical and software-based robots.
20   DNA stands for deoxyribonucleic acid. The DNA molecule carries the genetic information in human beings.

CHAPTER 1: ANALOG TO DIGITAL LIFE

21   Uber is a transportation network company (TNC) that connects cab service providers and consumers through a mobile app. Uber also allows ride-sharing at a lower rate.
22   Lyft is similar to Uber in that it has a mobile app that a person can use to order a cab. It also allows ride-sharing at a lower rate.
23   Google is a famous search engine that is available in the internet. Google is owned by Alphabet company.
24   Bing is a search engine offered by Microsoft that is available in the internet.

25 Siri in an application that runs on Apple iPhones. It can converse with humans and provide information, such as location of a specific restaurant or a list of nearby coffee shops.
26 The Apple iPhone is a mobile phone made by the same company that manufactures Apple computers.
27 Google Maps is a mobile app that provides directions to drivers or pedestrians by using Global Positioning System or GPS.
28 Waze is a mobile app that provides directions to drivers or pedestrians by using GPS.
29 Bluetooth is a wireless protocol for sending messages in short distances between devices made by different manufacturers.
30 Facebook is a social media platform for sharing words, photos, and videos with friends.
31 Instagram is another social media platform for sharing words, pictures, and videos. Instagram was bought by Facebook in 2012.
32 Twitter is a social media platform that allows registered members to broadcast short "tweets."
33 Webex is an audio and video conference call facility. Participants dial in to a "bridge" number and can conduct a group discussion and even share or display files for all participants to view.
34 Zoom is an audio and video conference call facility. Participants dial in to a "bridge" number and can conduct a group discussion and even share or display files for all participants to view.

CHAPTER 2: THE BIRTH OF ARTIFICIAL INTELLIGENCE

35 Link to an article on the Harvard University website on the history of artificial intelligence: http://sitn.hms.harvard.edu/flash/2017/history-artificial-intelligence/.
36 Link to a Wikipedia article on AI winter: https://en.wikipedia.org/wiki/AI_winter.
37 Link to a *Time* magazine article regarding the inventor of the first microprocessor: http://content.time.com/time/magazine/article/0,9171,155487,00.html.
38 CPU stands for central processing unit, which is responsible for a computer's arithmetic and logic operations.
39 Link to the MPT4U website: http://www.mpt4u.com/.

40 Link to an article on the SAS website regarding the history of natural language processing: https://www.sas.com/en_us/insights/analytics/what-is-natural-language-processing-nlp.html.
41 Robotic process automation or RPA is a subdomain in AI involved in automating logic-based and rules-driven functions of humans. Link to a McKinsey&Company website on RPA: https://www.mckinsey.com/business-functions/digital-mckinsey/our-insights/the-next-acronym-you-need-to-know-about-rpa.

CHAPTER 3: VALUE PROPOSITION

42 Link to the Blue Prism website: https://www.blueprism.com/get-started?utm_campaign=BP_NAM_US_EN_Brand_SEM&utm_source=adwords&utm_medium=cpc&utm_adgroup=65543612385&utm_content=343958344289&utm_term=blue%20prism&device=c&position=1t1&gclid=EAIaIQobChMIz6v7lPTj4QIVsR-tBh180A-DEAAYASAAEgL0zvD_BwE&gclsrc=aw.ds

CHAPTER 4: HOMEWARD BOUND

43 Here is the link to the YourTube video on Google Duplex: https://www.youtube.com/watch?v=D5VN56jQMWM
44 Link to the *Forbes* article: https://www.forbes.com/sites/eustaciahuen/2016/10/31/the-worlds-first-home-robotic-chef-can-cook-over-100-meals/#638a5327228b.

CHAPTER 6: EMOTIONAL INTELLIGENCE

45 Sentiment analysis is the process of mathematically calculating the sentiment of the writer of a text, whether it is negative, positive, or neutral.
46 Link to the IBM Tone Analyzer API web page: https://cloud.ibm.com/apidocs/tone-analyzer
47 Link to a web page on Bitext Text and Sentiment Analysis API: https://rapidapi.com/bitext/api/bitext-text-and-sentiment-analysis?utm_source=mashape&utm_medium=301

48 Link to Twinword's website: https://www.twinword.com/api/emotion-analysis.php
49 Link to a Brandwatch web page on sentiment analysis: https://www.brandwatch.com/blog/get-a-deeper-understanding-of-consumer-sentiment-with-emotion-analysis/
50 Link to Hanson Robotics' website: https://www.hansonrobotics.com/

CHAPTER 7: THE ART OF MAKING ART

51 Link to goodreads.com, the source of the quote: https://www.goodreads.com/author/quotes/115200.Johann_Sebastian_Bach.
52 Copyright is the legal right to control the reproduction and selling of music, books, films, photographs, or plays.
53 Royalties refer to the sum of money paid to a person for the use, sale, or performance of a material copyrighted to that person.
54 Here is a link to the Futurism.com article: https://futurism.com/the-worlds-first-album-composed-and-produced-by-an-ai-has-been-unveiled
55 Link to an article on christies.com covering the sale of an AI-generated artwork: https://www.christies.com/features/A-collaboration-between-two-artists-one-human-one-a-machine-9332-1.aspx.
56 Link to a webpage on the *Encyclopedia Britannica* website containing a write-up on Beethoven's Symphony No. 9: https://www.britannica.com/topic/Symphony-No-9-in-D-Minor.
57 Link to a webpage on the *Encyclopedia Britannica* website containing a write-up on Chopin: https://www.britannica.com/biography/Frederic-Chopin.
58 Link to a webpage on the *Encyclopedia Britannica* website containing a write-up on Impressionism: https://www.britannica.com/art/Impressionism-art.
59 Link to the pablopicasso.org website containing an article on the master: https://www.pablopicasso.org/blue-period.jsp.
60 Link to an article on accademia.org about the life of the master artist Michelangelo: http://www.accademia.org/michelangelo/.

CHAPTER 9: MINDING MY OWN BUSINESS

61 Link to the *New Your Times* article: https://www.nytimes.com/2019/02/04/business/family-tree-dna-fbi.html.

62 Link to BioCatch's website: https://www.biocatch.com/continuous-authentication-solutions
63 Link to the *Los Angeles Times* article: https://www.latimes.com/business/technology/la-fi-tn-microchip-employees-20170403-story.html

CHAPTER 10: WHAT ABOUT MY JOB?

64 Link to the Statista website on elementary schools in the United States: https://www.statista.com/topics/1733/elementary-schools-in-the-us/.
65 Link to the CNN article: https://www.cnn.com/2018/11/01/health/virtual-reality-education/index.html.

CHAPTER 11: TEACH YOUR CHILDREN

66 Link to the Grammarly website: https://www.grammarly.com/?q=brand&utm_source=google&utm_medium=cpc&utm_campaign=brand_f1&utm_content=329885936576&utm_term=grammarly&matchtype=e&placement=&network=g&gclid=EAIaIQobChMIn4Pj8T14AIVAlcNCh0WgAxiEAAYASAAEgK3pvD_BwE.
67 Link to the Cogito website: https://www.cogitocorp.com/.
68 Link to www.crunchbase.com, which contains details on Cogito Corporation: https://www.crunchbase.com/organization/cogito-corp#section-overview.

CHAPTER 13: YOU GOT A FRIEND IN ME

69 Link to Boston Dynamics website on Atlas: https://www.bostondynamics.com/atlas
70 Link to a write-up on the social robot named Sophia on the Hanson Robotics website: https://www.hansonrobotics.com/sophia/.

CHAPTER 14: REMEMBER MASLOW

71 Link to the Dell Technologies website: https://www.delltechnologies.com/en-us/perspectives/how-robots-and-ai-are-transforming-the-food-industry/.

72 Link to the Boston Dynamics news article: https://www.bostondynamics.com/press-release-boston-dynamics-kimena-pick-2019-04-02

CHAPTER 15: THE SCIENTIFIC WAY

73 Link to the Genetic Engineering & Biotechnology News website: https://www.genengnews.com/topics/drug-discovery/sanofi-exscientia-ink-up-to-e250m-deal-for-bispecific-drugs-against-metabolic-diseases/
74 Link to the Nobel Prize website: https://www.nobelprize.org/prizes/medicine/2009/press-release/.
75 Link to the *Digital Trends* article: https://www.digitaltrends.com/cool-tech/johns-hopkins-smart-stethoscope-ai/.
76 Link to Galen Robotics website: http://www.galenrobotics.com/about-galen-robotics/
77 Link to a Wikipedia article on remote surgery: https://en.wikipedia.org/wiki/Remote_surgery.
78 Link to an article on diagnosticimaging.com entitled "How AI Is Evolving in Diagnostic Imaging": https://www.diagnosticimaging.com/automation/how-ai-evolving-diagnostic-imaging.
79 Link to the Numenta website: https://numenta.com/.
80 Link to the *MIT Technology Review* article: https://www.technologyreview.com/s/511106/numentas-brain-inspired-software-adds-smarts-to-the-grid/.

CHAPTER 16: JUSTICE DELAYED IS JUSTICE DENIED

81 Link to the US Department of State website on the Freedom of Information Act: https://foia.state.gov/.

CHAPTER 17: A MIGHTY FORTRESS

82 Link to the Statista web page showing statistics on data breaches: https://www.statista.com/statistics/273550/data-breaches-recorded-in-the-united-states-by-number-of-breaches-and-records-exposed/.

CHAPTER 18: WHAT MAKES THE WORLD GO 'ROUND

83 Link to a *US News* web page regarding 200,000 new jobs reported for January 2018, indicating a strong labor market: https://www.usnews.com/news/economy/articles/2018-02-02/economy-adds-200-000-jobs-in-january-as-unemployment-holds-steady.
84 Link to the *Bloomberg* article: https://www.bloomberg.com/news/articles/2017-06-28/ai-seen-adding-15-7-trillion-as-game-changer-for-global-economy.
85 Link to an article on wired.com entitled "Why Man + Machine Is Better than Man or Machine for Investing": https://www.wired.com/brandlab/2017/10/man-machine-better-man-machine-investing/.

CHAPTER 19: THE NEXT FRONTIER

86 Link to an article on *Johns Hopkins Health Review* website entitled "Mind-Powered Prosthetic Limb": https://www.johnshopkinshealthreview.com/issues/fall-winter-2015/articles/mind-powered-prosthetic-limb.
87 Link to the Numenta website: https://numenta.com/.
88 Link to an article on Ray Kurzweil discussing singularity: https://futurism.com/kurzweil-claims-that-the-singularity-will-happen-by-2045.

ABOUT THE AUTHOR

Elzar Simon has more than thirty years of experience as an information technology manager at the global and regional level. He has lived in Australia, North America, and Asia and has worked as a senior IT leader and strategist in a broad range of industries, including finance, logistics, health care, government, IT services, and higher education. He currently works as a senior IT director for one of the largest and most prestigious universities in the Northeast of the United States.

CPSIA information can be obtained
at www.ICGtesting.com
Printed in the USA
FSHW022015181119
64242FS

9 781480 878358